M000201744

THE MOBILE WAVE

THE MOBILE
WAVE

How Mobile Intelligence
Will Change Everything

MICHAEL SAYLOR

DA CAPO PRESS
A Member of the Perseus Books Group

Cataloging-in-Publication data for this book is available from the Library of Congress.

First Vanguard Press edition 2012
First Da Capo Press paperback edition 2013

ISBN 978-1-59315-720-3 (hardcover)
ISBN 978-0-306-82253-7 (paperback)
ISBN 978-1-59315-721-0 (e-book)

Published by Da Capo Press
A Member of the Perseus Books Group
www.dacapopress.com

Da Capo Press books are available at special discounts for bulk purchases in the U.S. by corporations, institutions, and other organizations. For more information, please contact the Special Markets Department at the Perseus Books Group, 2300 Chestnut Street, Suite 200, Philadelphia, PA 19103, or call (800) 810-4145, ext. 5000, or e-mail special.markets@perseusbooks.com.

10 9 8 7 6 5 4 3 2

To my parents,
Phyllis Ann and William J. Saylor, and Carl Kaysen,
who inspired my passion for the history of science.

C O N T E N T S

FOREWORD

I'm a technologist. And technologists, by our nature, embrace change, radical change. The technological wave marking this second decade of the twenty-first century is indeed radical. It is disruptive and transformational. But there is no need to fear it. We, as individuals, must learn to understand its current state and its future potential to affect our daily lives; leaders of corporations and heads of governments, as well, must adapt to its powerful forces or risk seeing what they have built perish in a relative instant.

I've written *The Mobile Wave* with the kind of appreciation for today's formidable technological currents that a veteran sea captain or sailor might have for the Deep Blue and the rogue waves that can suddenly appear in its midst. Understand the wave, you can ride it. Refuse to adjust, you will be swallowed. We have seen, in a short amount of time, the disappearance of a large number of household brands that failed to take sufficient and early heed of the software revolution that is upending traditional bricks-and-mortar businesses and creating a globally pervasive digital economy.

Products we've lived with our entire lives are becoming software because of mobile computing technology. Magazines and newspapers are becoming software, as are books. Everything in our wallets from IDs to cash will become software, as we adopt pay-with-your-phone technology. Music, for nearly a decade, has been software, as Steve Jobs and the team at Apple so thoroughly understood. We see this transformation of physical objects into software rippling through industry after industry, driving changes like the bankruptcy of Kodak and Borders Books, which were rooted in the physical world, and the ascendancy of Amazon, which is rooted in the cyber world.

What amplifies the transformational power ahead is the confluence of two major technological currents today: the universal access to mobile computing and the pervasive use of social networks. Social networks radically increase the use of computing devices, and mobile computing increases the usefulness of social software. It's a virtuous cycle that magnifies the impact of both waves.

It's easy to underestimate the power of information. Mobile technology puts real time information in your pocket, allowing everyone to magnify his or her knowledge in any setting. In the hands of business executives we see faster and smarter decision making. In the hands of consumers we see smarter buying via "hijacked retail." In the hands of third world farmers we see much more efficient markets. And real time information distributed through growing mobile social networks has proven powerful enough to drive revolutions, toppling long-standing governments in a matter of days. Information is powerful, and we are living through an Information Revolution, with consequences comparable to the Agricultural and Industrial Revolutions.

Developing nations are leapfrogging into the twenty-first century with the help of smartphones and tablets and cell towers. I firmly believe this will be a game-changer for the global economy: the ability to deliver a First World education—as well as critical, time-sensitive information—to nearly everyone for a thousand times less capital. By 2015, at the rate the world is building and consuming smartphones, we're going to have 4.5 billion such devices connecting people worldwide. Mobile communications can only improve the quality of life for most, particularly in those parts of the world where paved roads and crowded airports are things left to the imagination. This is incredibly exciting for all of us who are interested in progress in general.

It is not just a question of economic betterment. Technological forces today—in the form of social networks such as Facebook and

Twitter—are offering a chance at overnight democratization, a leveling of entrenched power in some of the most authoritarian states on record. These software application networks are giving new constituencies a voice. They will continue to destabilize many traditional institutions, as we have seen in the remarkable Arab Spring uprisings. Let there be no doubt: these same mobile networks and social networks will fundamentally impact politics in Europe, the United States, the Middle East, Asia, Latin America, and Africa.

These are forces that combine to form the Mobile Wave, a surging tidal force bearing down on corporations, governments, universities, nonprofits, and nearly every facet of society. These fundamental forces can either be harnessed for good or bad. As a founder and CEO of a large publicly traded company, I can assure you that they can be harnessed to help you grow your brand and build your business. They, alternatively, can be ignored and be harnessed by someone else—someone out to tear down your brand and destroy your business.

I personally believe that the combined forces of mobile and social software networks will transform 50 percent of the world's GDP in the coming decade. They are reaching a crescendo level that will remake businesses, industries, and entire economies. With this book, it's my hope to provide some navigational insights so that we may all learn to ride the wave upon us, to harness the power of information technology, and to come out on top.

CHAPTER 1

THE WAVE

The Shape of the Wave

Introduction – Mobile Computing Is Different, Not Just Smaller – The Universal Computer – Replacing Physical Products with Software – Planting the Software on Your Customer – The Changing Nature of Software: From Solid to Vapor – Disruption Ahead – Operating Businesses in Cyberspace – Accelerating the Information Revolution

In late June 2010, I happened to be eating lunch at New York's Blue Water Grill with my niece Lauren and her mother. Lauren was turning thirteen, so I had flown them up from Sarasota, Florida, and shown them the town.

"Well, Lauren, what do you think of the Big Apple?" I asked as we settled in.

"I really like the operating system," she replied, "and I want one for my birthday."

It took a few moments, but realization dawned. To her generation, Apple was a company, and the "Big Apple" was the iPad—not the city. I knew teenagers liked iPhones, but the iPad had only been available for three months, and the eagerness she showed for it came as a surprise.

"Why do you want an iPad?" I asked, leaning forward.

"I can do my schoolwork on it, and I can play games on it," she said. "I can read my email on it. And all my friends have one."

1

"How many of your friends?" I asked.

"Hundreds," she insisted. For me, this was an epiphany. Already her circle had snapped up the twelve-week-old tablet like candy.

I had already come to believe that businesses would be keenly interested in tablet computers because of their ability to provide information anywhere and anytime to their workers. But suddenly, I realized that this same device was likely to transform the consumer space as well. The confluence of business interest and consumer interest would create a virtuous cycle of adoption, lower cost, and application innovation.

Whenever teenage girls and corporate CEOs covet the same new technology, something extraordinary is happening.

Over the following weeks, I had a series of additional telling observations. Once, during a walk along the beach in St. Tropez, I spotted a child, probably three years old, in a stroller. He was playing with an iPad. A Swedish woman leaned over and spoke to him.

"What are you doing?" she asked.

"I'm working on my music," he said.

Such a thing would never have occurred with a laptop computer! And I doubt this three year old was some kind of prodigy. I think his parents were simply early adopters of what will become an essential learning tool for the next generation. Indeed, if you browse through the Apple App Store today, you'll find a wide array of learning and creativity applications that are more approachable than anything ever produced before.

Watching that three-year-old boy, I realized just how envious I was of this young generation. I would have loved to write music or build skyscrapers with the unlimited resources available in cyberspace, rather than with the xylophone and wood blocks my generation had available to it. One might think $499 is too much money for a toddler's toy, but consider how much families spend on toys, books, and videos. In addition to creating music and architectural wonders, this

$499 investment would also enable the youngster to read books, sing songs, watch movies, and play games, and it would even help parents to monitor their child, as needed. With new software and updates continually becoming available, the iPad will become the obvious platform for many new birthday and holiday presents.

Suddenly, $499 seemed less like an extravagance and more like a necessary replacement for older, out-of-date equivalents. Games, toys, music, books, and even nannies are being transformed into software, and this new device will become the universal container for all of them. And the adult population will be impacted just as much, and perhaps more, as many physical objects we find in our pockets and purses today—like keys, wallets, credit cards, calendars, cameras, recorders, maps, and mirrors—become software, too, thanks to the mobile computing technology.

I also heard from a friend who bought his seventy-year-old mother an iPad. She has never been able to use a PC very well. The mechanics of the mouse and the bewildering array of interactivity on websites were too great a barrier. Infirmities prevented her from spending much time sitting upright in a chair, staring at a monitor. But she could easily sit on a couch with a tablet cradled in her lap.

Now she receives pictures every day from her children, who snap them on their cell phones and email them to her. She does her banking and shopping using mobile apps because they are far simpler than websites, and she watches her favorite television shows "on demand." Once she learns how to use FaceTime or Skype for video connections, she will be able to read bedtime stories to her distant grandchildren and virtually attend birthday parties thousands of miles away.

So here was a new technology that appeals to three-year-old toddlers for play and learning, to thirteen-year-old girls for games and social interaction, to fifty-year-old CEOs for real-time information and enhanced decision-making, and to seventy-year-old grandmothers to stay connected with their children and grandchildren. All in a fraction of a year since its introduction.

But the effects of mobile technology reach far beyond the personal level, to have a startling impact on society.

For decades, rulers such as Hosni Mubarak of Egypt and Muammar Qaddafi of Libya had cowed their citizens with censorship, spies, and military strength. But in the so-called Arab Spring of 2011, people defied the national muzzle. Revolts flared up and were sustained by real-time social media and other platforms enabled by mobile devices. Such communication power in the hands of large numbers of individuals could not be suppressed, and today Mubarak and Qaddafi no longer reign. Though the ultimate winners and the new equilibrium remain unclear, the tumult was unprecedented, and *all* governments have come to realize that there is a fundamental new dynamic afoot.

Mobile Computing Is Different, Not Just Smaller

It's easy to fall into the trap of assuming that a new technology is very similar to its predecessors. A new technology is often perceived as the linear extension of the previous one, and this leads us to believe the new technology will fill the same roles—just a little faster or a little smaller or a little lighter.

Yet every now and again, a truly *disruptive* technology appears and causes major changes to business, society, or economies. It yields nonlinear effects on so many levels and at such a grand scale that it's very hard to grasp the scope until after the dust settles.

Mobile computing is this type of disruptive technology.

Mobile computing encompasses tablet computers and the newest generation of cell phones that I will call "app-phones" in order to distinguish them from previous generations of smartphones and feature phones. Those earlier models ran a limited range of applications and enabled users to browse the web. App-phones are true computers, with operating systems that are capable of supporting a wide range of applications and programming languages. In fact, they should be thought of as computers first and phones second.

But if app-phones and tablets were just small computers, they wouldn't have much impact. We had small portable computers in the form of Windows-based "handhelds" and "pocket PCs" as early as the late '90s and early 2000s, and they didn't have much impact. So why should we think that small phone-computers and tablet-computers will be much different?

It turns out that the new mobile computing technology brings much more than phone calling to small computers. It also brings a dramatic new model for applications (the app), a whole new ecosystem in which to create and distribute those apps (app stores), and a whole new user interface experience (multi-touch). They are the least expensive computers on earth with the lowest-cost applications, making them affordable—for the first time—to the majority of consumers. And because most mobile devices are also phones, they will be the first computers to become truly pervasive, in most every pocket, purse, or backpack.

The Universal Computer

Mobile computers will become the standard universal computing platform on the planet.

As of 2011, more than 5.3 billion people in the world possessed cell phones.[1] That's about 70 percent of the earth's population. The use of traditional cell phones is waning, and most of these people will upgrade to full app-phone capability in the coming years. That means more than 5 billion people will soon be carrying a computer in their pockets. The sheer scale of these numbers will cause mobile computing to become the standard universal channel through which people will receive services from businesses and governments, and through which they will interact with one another using social media.

Currently people ask, "Why do I need a tablet computer or an app-phone to access the Internet if I already own a much more powerful laptop computer?" Before long the question will be, "Why do I need a laptop computer at home if I have a mobile computer that I use in every aspect of my daily life?" In 2010, 42 percent of Americans surveyed said they "can't live" without their cell phones,[2] and two thirds slept with their cell phones next to their beds.

You will never see those statistics with laptop computers.

Greater computing power does matter, but it matters a lot less than pervasive access to computing power and availability of a wider range of applications.

No other computing technology can beat the 24/7 accessibility of mobile technology. No other computing technology can beat the wide range and low price of applications available for mobile computers, and the supply of those apps will explode. In the mobile computing world, a developer could create a $1 app that might be used by five billion people. That's economic incentive!

This new mobile computing platform will cause two fundamental changes to how businesses use software in the conduct of their business. It will cause companies to replace their physical products and services with software equivalents, and it will cause companies to extend their business processes beyond the four walls of the business and out to the software resident on their consumers' mobile computers.

Replacing Physical Products with Software

Mobile computing will cause many companies to replace their physical products and services with software equivalents. The software versions of products and services will embody radical new features that are not possible with their physical counterparts. And the software-based products and services will be inherently cheaper to create. Factories will not be needed, distribution networks will be less necessary, and brick-and-mortar stores will disappear. Vast amounts of capital will be driven out of the production equation.

We've had a taste of this upheaval in the music industry. In the physical-product world, if you decided to buy a specific song, you drove to a record store, pored through the racks, and paid $16 for a CD that was physically packaged with a back-load of songs you probably didn't actually want. To get the engine you had to buy the train.

Then MP3 and the iPod came along and turned music into software. As software, music no longer had to obey the physical constraints imposed by physical media. Thus it became possible to distribute music electronically (via iTunes), sell music by the song (for $0.99), and market music directly by the artist, without the involvement of any middlemen.

The iTunes store wasn't a linear extrapolation of the record store model. It didn't just replicate the physical music packaging and sales model. It *reinvented* music. Between 2003 and 2007, more than 2,700 record stores vanished,[3] freeing up real estate and capital that could be used for other things. And thanks to the inexpensive nature and ease of the modern music purchase, consumers have more money to spend on other things.

Music is just one example of the new business models spawned by the shift to software on mobile devices—there are others in various stages of evolution. Publishing is turning into software. Travel agents have turned into software. Advertising is becoming software. News is becoming software. The credit and payment industries are under attack by software-based rivals like PayPal and many new start-ups who want a piece of the billion-dollar money stream currently going to the credit card companies. The software revolution is happening all around us.

Because software-based products and services require so much less capital to manufacture and sell, there will be an explosion of start-up companies that will enter markets that previously had high barriers to entry. These start-ups will be hungry, aggressive, and innovative.

With the parallel rise of cloud-based computing services, these start-ups can avoid even the capital they once needed for data centers to run their new products and services. They can grow their computing capacity incrementally, and at pace with the uptake of their products and services. It's a remarkable low-barrier formula for innovation and commercialization of new products, and one that will be tapped by start-ups around the globe.

This is the start of a golden era for business-savvy twenty-something software engineers with great imagination and nothing to lose.

Planting the Software on Your Customer

Mobile computing will force every consumer-facing company to establish a direct application linkage with its customers.

Given the ability to plant their software directly on the person of every customer who owns a mobile device, businesses will open an explosive new front in the battle of competition. They will extend their business processes to include the customer directly in the sales process, influence and advise customers in a more intimate way, deliver value-added services to better differentiate their products, and interrupt the customer's relationship with competitors. This direct application connection will shorten the value chain from manufacturer to consumer, cutting out middleman operations of retail, distribution, and service. Consumers should enjoy lower prices made possible by the elimination of the middleman costs and enjoy more direct and personalized service from their preferred suppliers.

With a compelling personalized mobile application, a company will be able to achieve a one-to-one marketing relationship with each of its customers that can influence buying decisions that already are under way—even when those buying decisions are taking place in a competitor's store! Many of the big box retailers have already seen this. Consumers browse the brick-and-mortar store, examine the products, identify their favorite, and use an app on their mobile phones to scan the barcodes and identify an online retailer with a better price—and then the consumer will have the product shipped directly to their homes.

This is "hijacked retail." One retailer hijacks the sale of another retailer simply by inserting himself into the sales process through their mobile app.

Every company will quickly recognize the potential benefit—and inherent danger—of this new connection. There will be a scramble as companies try to figure out how to offer special deals, discounts, loyalty programs, virtual sales assistance, product advice, and other services directly through their apps on their consumers' mobile devices. The goal will be to offer the most useful app in your industry so that your app captures the attention, the usage, and the loyalty of the consumer base first.

First movers will take and hold a disproportionate share of a consumer app usage—especially in the world of social networking and viral references. Whereas in the 1990s companies scrambled to acquire the best Internet domain names, today the winners will be those companies whose app icons appear on the first two pages of people's phone screens.

The Changing Nature of Software—from Solid to Vapor

Mobile technology is changing the nature of software.

It's causing software to transform from a "solid form" to a "vapor form." With desktop computers, software exists in a solid form—like a large rock that sits on a desk, to which people must go to use the software. That imposes severe restrictions in when and where people can use their software, but lacking anything else, we've been very happy to have those rocks.

With laptop computing, software exists in "liquid form." It's available along river ways and oases demarcated by coffee houses where people can sit at the Wi-Fi watering holes.

With mobile technology, software has boiled off into a vapor form that surrounds us everywhere we go. Unlike the visits to the rock or the drinks at the watering holes, we will use vaporous mobile software as constantly as we breathe.

To fully grasp the new gas-like nature of mobile software, visit a shopping mall and observe a group of teenagers. In a group of four, you might see one person talking, one listening, one texting, and the fourth connected to the Internet on her phone. Yet these teenagers will all be engaged in the same conversation. The texter is extending the conversation to distant friends. The one on the Internet is posting the conversation to a Facebook page, as well as checking the time of the movie they intend to see. These teenagers are breathing the mobile technology that is available to them. They are using it as an integral part of their immediate social activity.

Alternatively, observe a group of business people engaged in a meeting. In the traditional model, each person will be talking or listening, and when a topic comes up that requires outside information, someone will say, "I'll get back to you on that," or "I'll email him later and find out." The decision-making slows down or grinds to a halt. With desktop computers, people get their information and communication only when they visit the rock. Even liquid software on a laptop is not a replacement for mobile software since it requires participants to be seated, with Internet access, a fully booted up computer, and superimposes a screen in between all participants.

But as the teenagers have shown us, mobile software can be used in every moment of daily activity, whenever and *wherever* it needs to be available.

Imagine what it will mean when decision-making information is always available during the course of any business conversation. Imagine discussions with employees where the latest production numbers are just a tap away. Or conversations with customers where their latest orders can be reviewed and prices can be compared in an instant. Or conversations with suppliers where years of performance data can be reviewed and analyzed from every angle.

Not only will business be a lot "smarter," it will be a lot faster, too.

Disruption Ahead

Mobile computing technology will cause software to replace physical objects and services. It will provide a universal computing platform to the majority of humankind, and it will spur the creation of innumerable new applications that are not possible without a universal networked computer that is carried on one's person 24-hours a day.

This new capability will disrupt long-standing behaviors and institutions affecting consumers, businesses, governments, and the global economy.

The Destruction of Paper: Paper has been the most common container of information on our planet. It holds novels, news items, magazine stories, homework assignments, and business reports of every type. But the mobile screen is magic paper. It can show any page and blend text with multimedia seamlessly. It's searchable, send-able, and zoomable. It weighs the same few ounces whether it holds one book or one thousand books. Why buy a physical book when you can hold the entire Library of Congress in your hand? Why pay for a map when a free one can speak to you? Soon the long-mocked "paperless office" will become everyone's office.

Instant Entertainment: Other containers such as DVD cases and film stock have held movies, television shows, video games, and photos. To enjoy their content, you had to accommodate their restrictions. You had to travel to where they were made available. Or you had to watch them according to a broadcaster's schedule. But on the mobile screen you can watch what you want, where you want, and when you want.

The Intelligent Wallet: Money, credit, and loyalty programs will all become software on your phone, taking a myriad of new shapes enabled by software. You will be able to give digital cash to your daughter, but limit the time window in which she can spend it. Or let your son use your credit account—but only in a few predetermined stores to buy school clothes. Digital cash knows who owns it, who should have it, and it can alert police if either of those is suspect. With digital cash, credit fraud could drop so sharply that companies could justify giving every patron a mobile device for free with the money saved.

A Showroom World: With mobile technology, we will be able to purchase any item we see in the world around us, instantly. If we see a neighbor's sofa that we like, we can order one instantly and have it shipped to our home from the lowest cost supplier. Our environment will become a megastore showroom, and the need for retail inventory and physical showrooms will diminish. The effect will be especially pronounced with big-ticket items like furniture, electronics, and cars, but it will impact low-end items, too.

Hyperfluid Social Networks: People spend far more time on social networks sites than on any other internet destination. Mobile technology will amplify that usage, making social interconnection instantaneous and pervasive. People will live in social telepathy with their friends. With more than 800 million subscribers at the start of 2012,[4] Facebook holds the world's richest repository of consumer demographic and psychographic data. Businesses will tap into that data to create an array of new "friendly applications" that tie businesses to consumers in more rewarding and loyal relationships.

Worldwide Availability of Medical Care: Genuine globalization will occur when it will be possible to hire a doctor in Bangalore to examine us through our mobile screen for $10. That physician will check our temperature, blood pressure, heart rate, and conduct an EKG through med sensors connected to our phones or available at a medical kiosk. Medical service centers, like call centers, will arise in places such as Brazil and China to serve the world, creating tremendous price competition. Mobile technology will also bring doctors' presence to sick farmers in isolated villages and will monitor populations for outbreaks of disease.

Universal Education: U.S. public education is costly, and it's less effective than education in many other modernized nations. But why should an education system incur the costs of 20,000 people to teach eighth grade algebra, if only 2,000 are excellent at it. Mobile technology can bring the nation's best teachers and top experts into every classroom and improve the quality of education while freeing-up budgets. Skilled engineers can be trained for the cost of a few hundred dollars a year instead tens of thousands. In developing nations, where approximately one-fourth of children never finish primary school, and one billion people remain illiterate,[5] mobile computing will spread education where it hasn't gone before.

Jumpstart to the Emerging World: Mobile technology is a stunning gift to economies in places like India and Africa, which can leapfrog older capital-intensive infrastructure and join the newer mobile-enabled world. In some areas, mobile devices will be the first true infrastructure that residents have ever seen, and already their economies are accelerating because of it. Perversely, developed countries find themselves at a disadvantage as they must replace entrenched infrastructures and institutions before they can take advantage of newer mobile-enabled technologies. The rise of developing countries through mobile infrastructure will remake global trade and human resourcing.

Operating Businesses in Cyberspace

"You may not be interested in war, but war is interested in you."

This quote is loosely attributed to Trotsky[6] and its idea applies to our situation today. Your business might not be interested in software, but software is interested in your business. Industries will be transformed by mobile software much more quickly and substantially than the Internet transformed them. Mobile technology will create new manifestations for common products and services, new mechanisms for delivery, new partnerships, new customer relationships, and new economics. In this software-based world, many businesses will make big mistakes by assuming the wrong "physics."

In our physical world, goods and services are subject to the forces we observe every day: gravity, air pressure, and friction. On earth, a pitcher can throw a baseball 3 feet in water and 300 feet in the air. We understand this and we design a game around these physics. If we were thrust into outer space, things would be very different. There's no gravitational field. There's no air pressure. There's no friction caused by ambient atmosphere. In space, a pitcher can throw a baseball 300 million miles, and if a base runner misses his slide into second by just a millimeter, he will careen on forever.

It's hard to imagine at first how one might design a game for baseball in outer space, but at least Newton's laws of physics still apply. So with a little thought, we could design an outer space version of baseball, and no one would be killed.

But in cyberspace, where software lives, there are no laws of physics to provide an anchor. There are no laws of conservation of mass, momentum, or energy. In cyberspace, a baseball can loop around the pitcher twice, then accelerate to the speed of light before turning into a beach ball that explodes into a thousand baseballs. Software-based products are constrained only by what the developers can imagine and by the physical things with which the software might ultimately interact.

Most companies that provide physical products and services today will make the mistake of trying to re-create those physical products in cyberspace. What they must do is re-imagine their products, services, and businesses altogether.

For example, a traditional lock company that is accustomed to the physical world might consider creating a software-based key for your cell phone that can open your home door. They might even extend the idea of their "software key" so that if you lost your phone, the company could deactivate your old key and download a replacement.

On the other hand, a software company wouldn't stop there—it would re-imagine keys altogether. It would create a key that I could send directly from my phone to another person's device, so a delivery person could get into my house to deliver a sofa while I was at work. But I could also specify that the key not be transferred to anyone else. It would only open the front door, and no interior doors. The key would only be valid for a two-hour time period that I stipulate, and then it would expire. And it would work only if none of my children were at home—something I could track by the use of *their* software keys. I could give keys to any of my neighbors, which they would use to watch my house while I was away, but none of those keys would work while I was home. And the variations can go on and on. Whose key would you buy—the traditional lock company's key or the software company's key? Obviously the challenge for all traditional companies that are accustomed to operating in the physical world will be to forget about physics and "think like a software company."

Traditional businesses that are suddenly thrust into the cyberspace with mobile software will be like the earthlings in science fiction books, who suddenly find themselves in outer space and get everything wrong. For instance, the executives in the newspaper business have given up on advertising, yet the executives in social networking are raking in tremendous profits. In 2011, Google and Facebook together were worth more than $300 billion.[7] They could buy every newspaper, every publisher, and every television company on the face of the earth. And somehow the traditional media companies seem unable to copy them. Why isn't the *Wall Street Journal* free, for instance, and why don't its executives expect to get 500 million users, so they can finance it with advertising? In physical space, they can't.

In cyberspace, they could.

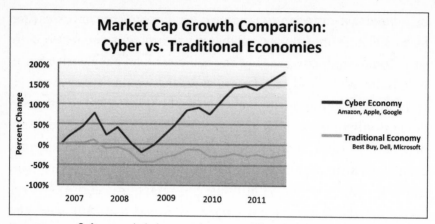

FIGURE 1.1 Software-minded companies are rapidly growing while bricks-and-mortar companies are declining.

Data Source: Ycharts, 2012.

In a friction-filled environment here on Earth, if I drop an item I need, it will roll downhill until it stops and I can walk over to pick it up. Drop the same item in outer space and I might have a split-second to grab it before it floats away and I lose it for eternity. Trivial events on the earth can become deadly mistakes in outer space and cyberspace. Companies that do well in cyberspace are ones like Google, Facebook, and Apple, because at the core they are software companies and they think like software companies—unconstrained by physics and scale. Those that do poorly are the bricks-and-mortar companies because their thinking is still bound by earthbound physics.

Many existing companies that fail to transform themselves into software companies and compete in the new cyber economy could disappear.

Accelerating the Information Revolution

There have been three great economic revolutions in human history—the Agricultural Revolution, the Industrial Revolution, and now the Information Revolution. Each great economic revolution was triggered by technology advancements that harnessed some form of energy, and by so doing, it freed up human energy that could be shifted to more varied and productive uses. The end result was an overall gain in worldwide wealth and better living standards.

In the Agricultural Revolution, humans harnessed bio-energy by domesticating plants and animals, thus freeing people from the daily burden of foraging and hunting for food. The resulting food surplus created a surplus in human energy that allowed hunter-gatherers to transform themselves into farmers, builders, craftsman, and shop-keepers. Nomads became city-dwellers, and everyone grew wealthier as the depth and complexity of their economies increased.

Likewise, with the Industrial Revolution people harnessed chemical energy from coal and oil to replace animal power, and in so doing opened the door for the manufacture of innumerable new products on a mass scale. Electricity then propelled the lumbering industrial complex past the limits of petrochemical fuels and steam-driven devices. Along the way, human energy was channeled from food production to product production and transportation systems, increasing the availability of physical goods at lower costs and increasing the overall wealth of mankind.

The Information Revolution is also about harnessing energy. In this case it's the "information energy" that makes everything we do in our economy more productive and more efficient. Computers are at the heart of the Information Revolution, helping us track and manage the assets of the world and automating many of the services that currently are labor-intensive, time-intensive, and error prone. Mobile computing, with its universal platform, will be the force that propels the Information Revolution to a dramatic new level. Just as electricity was the tipping point technology that accelerated the lumbering Industrial Revolution past the limits of petrochemical fuels and steam-driven machines, mobile computing is the tipping point technology that will accelerate the Information Revolution past the limitations of traditional computing.

However, every revolution has also brought disruption to the social structure, political systems, and economics. The transition from nomadic lifestyles in the Agricultural Revolution forced human societies to learn to live in large communities and to cooperate economically. It shifted the balance of economic power from the most talented hunters to the most talented planners. It shifted the political power from chiefs and tribes to kings and empires. In the Industrial Revolution, people left farms to live in densely populated cities. New lines of economic power were drawn between labor and management, and political power shifted from monocratic structures to more democratic structures.

And so, too, disruption will also be part of the Information Revolution in which we find ourselves today. It's early to say what form the disruptions will take, but it appears that privacy issues and social networking will drive new societal norms. Political power will be harder to hold by a dictatorial few. Brick-and-mortar establishments will be replaced by virtual storefronts. And labor will be displaced from the lower-skilled service workers as those jobs are automated away by software. As with all of the previous economic revolutions, the Information Revolution will free up human energy and capital that will be re-applied to the greater wealth and health of more people across the globe.

Mobile computing will be the most disruptive technology of our generation, and the revolution it leads is happening fast. The Agricultural Revolution took thousands of years to run its course. The Industrial Revolution required a few centuries. The Information Revolution, propelled by mobile technology will likely reshape our world on the order of decades. But despite the turbulence ahead, we live at one of the greatest times in history. Software will suffuse the planet, filling in every niche, and exciting opportunities will lie everywhere.

CHAPTER 2

COMPUTERS

The Evolution to Mobile Computing

The First Wave: The Mainframe – The Second Wave: The Minicomputer – The Third Wave: The Desktop – The Fourth Wave: The Internet PC – The Road Toward Mobile Computing – Early Mobile: From Cell to BlackBerry – The Fifth Wave: Mobile Internet – Multi-touch and More

A ninety-nine-year-old woman in Oregon bought an iPad as her first home computer. It changed her life, she said. It let her magnify text just by touching it, so she rediscovered one of her great lost joys: reading. She needed just the barest of instruction, since the screen felt like a playground, and she could use the iPad anywhere.

The iPad appealed to her partly because it distilled computers to a beautiful essence: just the screen. It didn't even seem to have circuitry. Yet it descended from screen-less machines that were as portable as a house and about as much fun as the Russian bureaucracy. Since then, computers have repeatedly shrunk, opened up to consumers, and conquered new environments.

Technologists have forged four great waves of computing, all leading up to mobile computing: the mainframe, the minicomputer, the personal computer, and the Internet PC. Each wave built on the previous one, and each had a greater impact on society.

The First Wave: The Mainframe

The Industrial Revolution extended human power in almost every domain. For instance, in the year 1700 a person needed 200 hours to make one pound of yarn; by 1824, the task took 80 minutes, freeing up more then 198 hours.[1] We became stronger by augmenting muscle power with motors. We became faster with cars. We extended our voices with the telephone. But no machine surfaced to make us smarter. The abacus and slide rule were centuries old, but they had only made math a bit faster. You certainly couldn't ask them questions and expect an answer.

The clever Charles Babbage invented a mechanical computer in the nineteenth century, which produced mathematical tables of logarithms and trigonometry. It existed, but it might as well not have. There was no business demand for it.

But World War II created urgent new demands for thinking machines. The Allies had to calculate the trajectory of artillery shells quickly and accurately in order to hit their targets. Yet the equations were maddeningly complex and a smart human needed at least twenty hours to "compute" them. Hence scientists developed the first modern electrical computer, the ENIAC (Electronic Numerical Integrator and Computer), which made the calculations in thirty seconds. Ironically, it had no effect on the war, since it entered operation three months after Hiroshima. But it led directly to the development of mainframe computers.

Mainframes were monsters.

The ENIAC weighed more than thirty tons and covered more than 1,800 square feet. As smaller machines were developed, they still filled whole rooms, cost millions, and could only be used by highly skilled operators. If you wanted it to solve a problem, you wrote a program, coded it in punch cards, put your stack of cards in line with other punch card programs, and waited sometimes days for the result, which arrived as a roll-like printout from a large electric typewriter. And if the program had a glitch—as it often did—you repeated these steps and waited again to see if you had fixed the error. Starfish interact more swiftly with the world.

At first, each mainframe was hand-built and unique. As the technology advanced, computers such as the UNIVAC (UNIVersal Automatic Computer) entered repeatable production, and became smaller and less costly. The UNIVAC found its first civilian use in the 1950 census, where it doubled the counting speed, and in 1952 it flouted pollsters by predicting the election of Dwight Eisenhower.

Mainframes let us perform feats that had previously been impossible, such as launch satellites and leave footsteps on the moon. But commercially their key contribution lay in automating the task of tabulating—eliminating the error-prone effort of repetitive calculations and record keeping. If a business needed to employ an army of people sitting at desks to carry out some exacting, repetitive task, that job was probably just right for mainframes. They took over billing, accounting, compound interest calculations, phone bill tabulation, inventory tracking, and paychecks.

The machines also streamlined logistics. Companies could handle a much greater flow of raw materials from truck to warehouse to factory, and the flow of finished goods from factory to warehouse to store. So items didn't linger in one place through lack of information, and companies could slash inventory costs. Airlines manage seats as inventory, and mainframes led to Sabre, a reservation system that let airlines broaden their routes and offer far more seats. Sabre cost American Airlines $40 million to implement in 1960,[2] an enormous sum at the time, but it easily paid for itself.

Mainframes enabled businesses to break through barriers of scale by automating the bookkeeping and other forms of tabulating that would have otherwise crushed operations. They enabled the multinational expansion of businesses in the 1950s and '60s. On a societal level, mainframes lowered costs, created new jobs, and yielded some new products. But mainframes didn't visibly affect the average person's daily life. We never saw them. We just heard about them and the army of lab-coated priests in horn-rimmed glasses who operated them.

Even so, mainframes were something new on the earth, and the public felt a kind of awe about them.

Mainframes were synonymous with computers through the 1960s because there were no others. By 1980, there were perhaps 100,000 mainframes sold. They were colossi with Rain Man brains, excelling at certain narrow tasks and not much else. You fed them immense tasks and then they went to work, slavishly counting millions or billions of records with tireless perfect precision.

The first mainframes used vacuum tubes as the switches for their logic circuits. The tubes were big and expensive and you needed a lot of them chained together to create the thinking logic of a computer. So early mainframes were expensive to build and delicate to operate, and they guzzled power. A single mainframe might use 160 kilowatts, enough to dim lights in a whole section of Philadelphia. Moreover, the tubes could overheat and burn out. As long as computers relied on vacuum tubes, these devices would remain costly rarities.

The wonder wasn't that they were smart; it was that they existed at all, especially given the cost.

Mobile computers and all modern computers owe their existence to the transistor. A team at Bell Labs developed this device in 1947—using paper clips and a razor blade for the early prototypes—and won a well-deserved Nobel Prize in 1953. The major innovation with the transistor was that it was made of semiconductor material like silicon, rather than expensive and fragile vacuum tubes. It was cool, fast, thrifty with power, and stable. And it was small. You could squeeze hundreds of these little three-legged spiders onto a small circuit board that previously would have taken a cabinet of vacuum tubes with its hundreds of yards of wiring. The vacuum tube was helpless next to this efficient invader and vanished after transistors entered mass production in 1955.

Transistors were originally produced as small discrete components, each one about the size of a pea with three stiff wires protruding from below. While this was much smaller than sausage-sized vacuum tubes, it still wasn't small enough to catalyze a new wave of computing technology. In fact, if they had tried to build the first IBM PC with discrete transistors, the 29,000 transistors in the Intel 8088 processor would have required a circuit board that was at least five feet on each side—assuming they were able to line up all the transistors in perfect rows and columns. If they had tried to build the venerable Intel Pentium microprocessor, which contains 140 *million* transistors, it would require a circuit board 390 feet on each side.

Clearly another technology innovation was needed to propel computing forward.

By the 1960s, technologists were perfecting a new technique for fabricating hundreds of transistors onto a single semiconductor "chip," rather than trying to wire up discrete transistors on a circuit board. This was the birth of the integrated circuit (IC). Today, integrated circuits can support billions of transistors on a single chip no bigger than the palm of your hand. This mass assemblage of circuitry opened the door for new, vastly less expensive computers, and many of the original mainframe providers—like RCA, Westinghouse, General Electric, Sylvania, and Raytheon—dropped out of the computing business, making room for computer start-ups that could enter with a lot less capital and new ideas for smaller computers.

The Second Wave: The Minicomputer

Scientists long chafed at the "priesthood" that guarded the mainframes, as they sought smaller, more accessible computers that would fit into a laboratory. Chips and IC's made possible the minicomputers.

"Minis" were about the size of a refrigerator and far cheaper than mainframes. Digital Equipment Corporation (DEC) introduced the first commercial success, the PDP-8, in 1965. At $18,000 a customer could buy five for the price of a small IBM 360 mainframe, and eventually the price would drop to $4,000. The low cost and smaller size of minis made possible many new applications that simply could not be cost-justified with mainframes.

The result was an explosion of new uses for computing.

Minis showed up on factory floors, where they controlled robots that assembled cars, cut glass, welded metal, sliced paper, and inserted parts into machines.[3] Soon they were running equipment everywhere from oil refineries to wastewater treatment plants. By the 1970s minis were acting as central servers for ticket agencies, running mass-transit systems, operating grocery store cash registers, processing radar data from CIA reconnaissance flights, and helping aircraft control towers all over the world. While mainframes automated the bookkeeping side of businesses, minis were automating the business processes themselves.

Along with DEC, new multibillion dollar companies sprang up to serve this market, boasting names like Data General, Prime Computer, Apollo, and Wang Laboratories. By 1979, minis were outselling mainframes 81,300 units to 7,300, and over a million had entered the world by 1985.[4] Part of the fascinating drama of the new minicomputer market was told in the Pulitzer Prize–winning book *The Soul of a New Machine* by Tracy Kidder (2000).

Meanwhile, somewhat out of sight, an innovation was emerging that soon would tip the scales of computing away from newly ascendant minicomputers. Microprocessors were being developed that would commoditize the very heart of the computer—the central processing unit, or CPU. While minicomputer and mainframe CPUs were highly proprietary electronics, jealously guarded by each computer manufacturer, microprocessors were produced as commodities, and sold to other manufacturers as a standard part.

Intel introduced the first commercially successful microprocessor, the Intel 4040, in 1975. While microprocessors were much less powerful than their minicomputer and mainframe CPU counterparts, they had the unique advantage of being available to anyone who wanted to make a computer. Suddenly clever twenty-year-olds in a garage could build their own computers.

And they did.

The Third Wave: The Desktop

In 1975 a cover story in *Popular Electronics* introduced an early desktop computer, the Altair, and it took many people by surprise. Suddenly, a small computer was sitting right in front of you. And you could own it. You could keep it at home. You could put it on your desk at work. You didn't have to share it with others. You didn't have to be a company, a scientist, or even a programmer.

The Altair was a build-it-yourself kit sold out of an Albuquerque strip mall, where a kid named Bill Gates was hanging around. Gates left Albuquerque, founded Microsoft, and developed the DOS operating system for the first IBM PC. The Intel 8080 microprocessor that powered the first IBM PCs was so important to success of the PC industry, that even to this day on the Microsoft campus you dial "8080" to reach reception.

In 1980, 500,000 personal computers were sold,[5] which is six times the number of minicomputers sold in the same year. And the PC industry was just getting started.

On the desktop, computing had entered another new world, and it quickly adapted to new uses. We had previously written using typewriters; now it was done using PC software. Tasks that required adding machines were now done with PC software. Information that had lived in file cabinets was stored using PC software, too. Even interoffice mail, once sent by messengers on rolling carts, was done by PC software. And in every case, the software was more powerful and more efficient than the physical mechanism it replaced. The days moved faster, more got done, and it was done better.

The PC automated the office worker.

It also led to innovations no one expected. One was the first "killer app"—the spreadsheet. The spreadsheet was an utter novelty when first introduced. What was *revolutionary* about the spreadsheet was that it was sort of like programming, but not. For the first time an average office worker could input a series of commands into the computer—such as, "Total up the numbers in cells A2 through A6 and add that sum to the number in cell D25"—without calling in a programmer. This capability unleashed a torrent of small and crucial spreadsheet applications to do things like manage finances, develop budgets, track inventory, handle call lists, perform "what-if" analyses, and much more. There appeared to be no end to what people could do with rows, columns, and a little math.

Almost overnight, the PC spreadsheet became an irreplaceable business tool.

The PC remade the office cubicle, and its proliferation has become the stuff of legends. By 1995 Bill Gates was the richest person on earth. Yet, throughout this era the often-anxious question continued to be asked: "Do I need a computer at home?"

In fact, not everyone did. But as we eased into the twenty-first century, that question seemed to evaporate. And the reason was the Internet.

The Fourth Wave: The Internet PC

The classic PC was self-sufficient. You bought programs for it at stores like Egghead Software and loaded them through the disk drive. You typed on it. You calculated on it. You played games on it. It was your own private tool for personal productivity and games. But the PC didn't stay isolated for long.

As early as 1960, J.C.R. Licklider—a psychologist-turned-computer scientist—proposed the idea of computer "thinking centers" and noted, "The picture readily enlarges itself into a network of such centers, connected to one another by wide-band communication lines and to individual users by leased-wire services."[6] He was describing the Internet before it was born. In his vision, PCs were portals into a much wider worldwide network of information. And since Licklider held a high position at the government's Advanced Research Projects Agency (ARPA), he possessed the influence needed to move the concept along.

ARPANET was born on October 29, 1969. Funded by the Department of Defense, it began as a research tool, but soon—to the surprise of many—it also became an active social network, engendering discussions of wine, science fiction, and other off-topic matters. The commercial Internet arose from it gradually. In 1979 CompuServe became the first company to provide email service to public PC users, and offered the first real-time chat in the following year. By 1996, countries around the world had connected into worldwide Internet.

If you could communicate across a computer network, you could do a great deal more. You could broadcast information, and news portals began appearing. You could barter and sell, and soon the world's biggest auction room emerged as eBay. You could sell retail goods and services, and the online store Amazon became better known than the great river that gave it its name. Using your own PC, you could search the entire Internet, and Google mushroomed. You could download music, and iTunes upended the music industry. Eventually you could broaden your social sphere, and sites such as Facebook appeared that developed a mass appeal akin to email. In the modern world, the question might still be asked: "Do I need a computer at home?" But today the answer would be distinctly different. You might not need it for personal productivity, but with every passing day it's becoming increasingly difficult to manage life without access to the Internet.

And for that you need a computer.

Moreover, the Internet has become central to society itself. Companies use it to process credit card transactions, for instance. It enables trillions of dollars of international trades every day, and it controls the electrical grid itself. We depend so much on it that if it were sabotaged, the United States might shut down for weeks.

As of 2011, most of the 1.2 billion PCs on earth are Internet PCs,[7] and the Internet itself is a vast jungle of billions of websites. We watch films and join online communities to hobnob with folks on the far side of the planet. We video-chat with friends, present webinars to colleagues, and offer our home movies to the entire human population. Every year, novel services arise on the Internet that satisfy old desires and introduce new ones we'd never imagined before.

If Gutenberg and his printing press created a galaxy of information, as Marshall McLuhan said, the Internet created a universe.

And, as mobile technology is showing, universes can expand.

The Road Toward Mobile Computing

For years, engineers were developing two major innovations that would bring forth true mobile computing: lithium-ion batteries and flash memory. These would provide the two elements necessary for the next great wave.

The Power Source: In the late 1970s, chemist Michael Whittingham of Binghamton University was experimenting with new kinds of batteries. He used the metal lithium as a pole in one of them and found that he had a potent energy source. But he also knew he had trouble.

As metals go, lithium is strange stuff. It can be cut with a knife. It floats on water, but it also reacts with water in an explosive way. And it's highly combustible, so it has to be packed in substances like petroleum jelly. In essence, it shouldn't be in the hands of millions of consumers. So researchers turned to batteries made of safer lithium-ion materials, which possessed the virtues of lithium without the hazards.

Sony released the first commercial lithium-ion battery in 1991. This battery was not only more powerful, but also significantly lighter than other rechargeable batteries. It could come in varying shapes and sizes, and so it fit into a wide assortment of devices. And the battery components were now safe in the hands of consumers.

Soon the United States military was relying on lithium-ion batteries, and in 2003 a surprise shortage of them reportedly almost halted the Gulf War.[8]

Portable Data Storage: Around 1980, Toshiba engineer Fujio Masuoka began looking for a way to store data on a chip that would work as well as a hard drive. Hard drives have almost every drawback imaginable for use in mobile devices. They are bulky, fragile, hot, and they drink up energy. The hard drive's one major strength is that it retains its data, even after the power is shut off, which is something chips could not do. Masuoka discovered a way to write and erase information quickly into non-volatile integrated circuits, and he called it flash memory. The name came from a colleague who said its erasure speed reminded him of a camera flash.

Chipmaker Intel recognized the potential of flash memory as far back as 1988, and flash memory helped drive the firm through an unprecedented growth spurt. Though commercially available, flash memory cost too much for common use. Yet as with any good technology, the price fell, its capacity grew, and its use spread. Consumers now use flash memory every time they stick a thumb drive into a USB port, and at trade shows people hand these out like candy.

Despite its many virtues, flash memory has shortcomings. It has a finite lifespan, and over time it slows done and errors increase. As the size of memory in a flash drive is increased, these problems worsen, and researchers are seeking better ways to manage their internal wear. Meanwhile, new and more sophisticated technologies wait in the wings, including phase-change memory, magnetoresistive RAM, and ferroelectric RAM.

Early Mobile: From Cell to BlackBerry

The first cell phones arose in the early 1970s from a heated race between Bell Labs and Motorola to develop the first portable handset.

Motorola's vice president Martin Cooper won in 1973, and used his device to phone his respected competitor at Bell. Just six years later, Nippon Telegraph and Telephone (NTT) of Japan unveiled the first "first generation" cell network (1G), which covered all of Tokyo's 20 million residents, and by 1984 access was available throughout Japan.[9]

The cell phone gave people a new capability: Two-way phone conversation from almost anywhere. But if the early cell phones had any visual displays at all, they just showed the ten digits of the number that had been typed into the device. The only "applications" were voice phone calling and perhaps voice mail. It wasn't possible to send a text message—there wasn't enough room.

The next major advancement was a screen that could show 24 or 48 characters. SMS (small message service), or text messaging, emerged with messages capped at 140 characters. As texting grew popular, Nokia rose to prominence. SMS ushered in the age of full keyboards, because once consumers were typing messages to friends, they wanted all of the characters at their fingertips.

The world's first smartphone, IBM's "Simon" Personal Communicator, entered production in 1993. Simon was like a Personal Digital Assistant (PDA) that had been merged with a cell phone. It had an on-screen keyboard and a simple touchscreen interface you controlled with your fingers or a stylus. The device was bulky, like all cell phones of the time, and it cost $899, or about as much as a PC. It stirred no frenzy in the marketplace—*PC World* said it "looked and felt like a brick"—but technology fails until it succeeds. Inventors had failed to fly airplanes hundreds of times before Kitty Hawk, and the Wright Brothers themselves tested some fifty different wing models in their shoebox-sized wind tunnel. The smartphone fit right into this pattern.

It failed and failed, and then it didn't.

The first camera smartphone arrived in 2002, with the Sony Ericsson P800. Suddenly people carried a camera whether they cared to or not, and found themselves using it more and more. Early images were blurry, almost as bad as the daguerreotypes of the 1800s, but they swiftly improved and phones escalated the attack on the film camera, already under assault from digital cameras. Also in 2002, people gained the ability to browse the web from their smartphones.

In 2007, GPS (global positioning system) joined the mix, with the Nokia N95.

The BlackBerry was the first smartphone blockbuster. It began as a two-way pager in 1999, and then evolved as a phone in 2002. It had a fan-like physical keyboard and a screen half the size of the device, excellent for texting and tolerable for short emails. By the end of 2009 its customer base swelled to 32 million,[10] many of whom became skilled "thumb-typists," and some executives became so addicted to it that it earned the name "CrackBerry." But it had little memory— the 2006 BlackBerry Pearl had 32 MB of flash, for instance—and it crashed all the time. The operating system was too weak.

Nevertheless, cell phones were looking less and less like phones, and more and more like computers. We were close to the true app-phone.

Few realized *how* close.

The Fifth Wave: Mobile Internet

With Steve Wozniak and Ronald Wayne, Steve Jobs co-founded Apple in 1978 and revolutionized computers in 1984 with the introduction of the Macintosh and its mouse interface. But in 1985 Apple's board of directors decided that Jobs wasn't "businesslike" enough, and ejected him. They replaced him with John Sculley, an executive from Pepsico. Sculley was businesslike, indeed, and Apple began a long, near-fatal tailspin.

Yet it was Sculley who introduced a predecessor to the iPad. In the mid-1980s Apple had begun developing a PDA it called the Newton. Newton MessagePad 100 debuted in 1993, and the little tablet included a stylus, handwriting recognition software, and built-in intelligence that would anticipate user behavior. However, it wasn't a phone, the interface wasn't engaging, and sales languished. Apple eventually scrapped it after ten years of development and nearly $100 million invested.

In 1997 a chastened Apple, facing serious financial difficulties, invited Jobs back as CEO. It was the smartest move the company had made since he left. In 2009 *Fortune* would declare him CEO of the decade,[11] and there were no other serious contenders.

In 2005, riding the success of the iPod, Jobs was worried. He had seen how camera phones had shriveled the digital camera market, and worried that phones with music players could undermine the iPod. He also realized that most cell phones were cheap, hard to use, and "brain dead."[12] So he set about creating a better one.

He introduced the iPhone at the Macworld Conference and Expo in January 2007 and released it the following June. Advance buzz was so glowing that some consumers camped out for days in front of Apple Stores, and hundreds lined up at every store in the nation. Top critics said the device warranted the buzz and described its interface with professionally tempered wonderment. The stores quickly ran out and enterprising folks on eBay and Craigslist offered iPhones for as much as $12,000.[13]

The iPhone offered a striking user experience. It was so small that about 300 iPhones could fit into the case of a single, less powerful iMac from 2000—an extraordinary leap in miniaturization. It was sleek, with a black glass face and a huge screen, twice as large as the BlackBerry's. Its 160 dpi resolution was double that of a traditional Macintosh and yielded gorgeous images. It had a revolutionary new touchscreen interface that eliminated keys and menus and let the user play with onscreen items like those in real life. It knew its physical orientation and flipped its screen from vertical to horizontal depending on how it was held—a startling experience at the time. And it had from four to eight gigabytes of flash memory, far more than the BlackBerry.

So applications that had never been possible on the BlackBerry exploded into prominence on the iPhone. Users could consult Google maps, watch YouTube videos, manipulate photos, and buy movies or airplane tickets. Apps appeared that took advantage of the motion sensing. Later versions offered GPS capability, so with the tap of the screen, data could be had concerning the area where the phone was located.

The BlackBerry lagged in all these areas.

As of late 2011, the iPhone operating system (called iOS) had taken more than 60 percent of smartphone and tablet market share.[14] Apple has maintained strict control over its operating system, but the technically savvy consumer can unlock it without approval, and get inside to add custom features. The process is called "jailbreaking." Apple has repeatedly tried to thwart it, yet jailbreaking occurs on all iOS devices, including the iPad, iPhone, and iPod Touch. As if in response to Apple's no-trespassing policy, in 2008 Google rolled out its Android series. Like the iPhone, the Android boasted a multi-touch screen, gyroscope, GPS, and web access. Unlike Apple, Google designed Android as an open-source platform, modifiable by programmers who might want it to behave differently than Google originally intended. Android can thus evolve based on the disparate ideas of a community of developers, whereas Apple continues to control tightly the direction and usage of its iOS operating system. Time will tell which approach will ultimately serve the needs of the consumer better and win in the marketplace.

In 2009, mobile Internet devices began outselling personal computers, 450 million to 306 million.[15] And in that year it became clear that Steve Jobs was going to introduce another mobile device—a tablet. Apple gave a public demonstration of it in January 2010, but unlike the iPhone, it faced skepticism before it appeared on the market.

"The unanswered question is whether we really need a 'third device'—something to fill the gap between smartphone and laptop," wrote two journalists in *PC World*. They noted that the public thus far had shown little appetite for tablets, and that the best tablet would be no more appealing than the best cod liver oil. "It's hard to imagine," they concluded, "all that many [people] will fork over the initial $499 for a crippled version, or as much as $829 (for the 64 GB/3G model you'd want)."[16]

Apple sold 16 million iPads in its first eight months, and Bernstein Research said the tablet was on track to become the fastest growing consumer product in history.[17] The journalists had failed to understand the fact that people weren't buying a tablet; they were buying an experience. And as important as the iPhone was, the iPad proved to be the real breakthrough.

The iPad used the same interface ideas as the iPhone, but its screen was far bigger, with much greater graphics capacity. These features opened it up. Users could now type on it effectively with the virtual keyboard. They could read books and magazines on it. They could work on it with productivity apps that the small screen had frustrated. They could immerse themselves in it. And it weighed only 1.5 pounds.

They could take it anywhere.

Multi-touch and More

I recall urging one man, "You should port your applications to the iPad, because it's so easy that even a three-year-old can use it. And without any training."

"Well, I'll go you one better," he replied. "It's so easy that my cat can use it."

Of course, his cat wasn't actually *using* the device, checking its email or playing Farmville. It just toyed with the screen. (Though this fact did suggest the possibility of a cat-friendly app—perhaps one that would let it go outside.) But the point remained: if an animal could manipulate a user interface, something remarkable had occurred.

The interface is crucial. If it's difficult, slow, or annoying, people will keep their distance. Fewer of them will use a difficult computer, and those who do will boot it up less often and explore it less fully. But when an interface is enjoyable, it will pull people in, and they'll explore its features in depth.

So the makers of early mobile computers faced a problem. They couldn't just move the PC interface over to a smaller screen that rested in your palm. The mouse wouldn't work, since it needs a surface, and even a stylus was clumsy, forcing the user to carry and employ a second object. No, the small screen required big innovation.

In 1971, physicist Sam Hurst was looking for a faster way to enter a huge pile of spectrometry readings into his computer. He came up with an early touch sensor, and by 1973 he had a much more refined device—one almost everyone would come to know. He placed two surfaces close together, in such a way that if he pushed one, it would contact the other and spark an electrical current between them. The pressure was a signal: input.

That's how ATMs work today, as well as voting machines, airport check-in stands, and museum kiosks. Hewlett-Packard released the earliest commercial touch-screen computer in 1983.

Pressure sensors remain the most common approach for touch-screens, but another uses infrared rays. Touch those screen and you break a tiny ray, creating a signal. The iPad employs yet a different method, in which the screen senses the electrical charge on your finger. You don't have to press anything, and a mere brush can become input.

But the true breakthrough with mobile wasn't touch. It was *multi-touch*. Instead of one finger controlling an action, two (or more) do, so you can squeeze and stretch the screen. Steps toward multi-touch began in the 1970s, even as the first PCs were appearing. In 1982, the University of Toronto's Input Research Group produced the first successful multi-touch system. When fingers touched a sheet of frosted glass, a camera behind it detected them as dark spots and registered each as a distinct input. The device could then respond. But what could be done with it?

In 1991 Pierre Wellner made the potential clear in his writings about a "digital desk" that supported multi-finger dragging and pinching. Various companies—most notably Apple—moved to develop his ideas.

Multi-touch, it turned out, was *fun*. Point-and-click had been fairly easy, but with the new tactile interface, the users entered an amusement park. They could spin, bounce, and slide screen objects. The whole interface seemed to exist on greased rollers, just below the screen. If you wanted to save an item for later, you could slide it out of view as you might in the physical world. You could even drag items off the screen itself and onto other devices connected with Bluetooth.

Multi-touch is enjoyable for another reason. It is *shape* processing rather than *symbol* processing, so it's wired deep in our minds. We comprehend shapes mainly with the visual cortex at the back of our brains, and this intricate skill is almost inseparable from sight itself. It feels effortless, because evolution has had millennia to automate it. The almost brainless crayfish can recognize the faces of opponents,[18] for instance, and you can teach bumblebees to favor one shape over another.[19] Shape processing is ancient and basic.

Symbol processing isn't. Only highly evolved creatures can perform it at all. Humans alone have the general-purpose capacity, with language, math, musical notation, metaphors, and symbols for almost everything. And though we find speech natural—we've done it for millions of years—reading can tax us.

"Reading is a bizarre skill—and a very complex process," said Harvard neuropsychologist Alfonso Caramazza.[20] We've been reading for a mere 5,000 years, so the brain has developed no areas dedicated to literacy. Instead, we hijack areas all over the brain, and as a result, a sizable portion of the populace is dyslexic.

In computers, the old interfaces required literacy, but multi-touch doesn't.

Moreover, shape processing appears earlier in life and lasts later, so eighteen-month-old infants can process with multi-touch, but most children can't really address a computer—in a traditional way—until they're seven or older.

The brain likes shapes.

Indeed, the difference between PC and multi-touch interfaces is staggering, if only for the ease with which people can perform simple tasks. To enlarge text on a PC, the ninety-nine-year-old woman in Oregon had to select the characters to enlarge, go to a text-size box in a menu bar, and choose a number representing her desired size. And then she might have to do it again, since it's often hard to predict the readability of a font at, say, 16 point versus 20 point. With an iPad, she just touches the screen and enlarges the distance between her fingers until text is the right size.

It's simpler. It's pleasanter. It's better.

Other factors make mobile devices very different from previous smartphones or laptop computers.

Widely Affordable: In 2011 an iPhone 3GS could be purchased for $49, a fraction of the price of a laptop. Moreover, mobile computers need none of the peripherals—no separate monitor, keyboard, mouse, speakers, modem, or yards of external cables—which increase the cost of setup to the high hundreds to mid-thousands. The devices also use cheap bandwidth. For $10 per month, owners could buy unrestricted access to the web, while they might pay $30–$80 per month to enable Internet access for the PC. So mobile is reaching blue-collar workers, children, seniors, and tillers in the emerging world. It will bring desktop capacities—and more—to people without desks and only moderate income.

Battery Life: In 2011, the iPad had a 24.8-watt hour lithium-ion polymer battery that can play continuous audio for almost six days. Apple claims it will carry the user through ten hours web surfing, videos, or music, and some have found that it lasts longer. With a laptop, by contrast, the battery commonly lasts for two to four hours of continual use. Just as cordless phones keep users tethered to a base charger, laptops keep their users hovering around wall sockets. This limitation affects the entire experience.

Instant-On: Mobile devices turn on instantly, and this is essential for social media usage for many people—especially teens, since many are always connected to their circle of friends. It's also the essence of business intelligence, where meetings need instant information. In fact, "instant-on" lies at the heart of people's relationship to their mobile devices. It separates a tool from a personal accessory.

Apps: Apps are software that drop into your mobile device from the ether. They are as cheap as candy, costing typically from one to ten dollars, are easy to buy, and many are free. Instead of paying hundreds of dollars for a word processor full of features whose names and functions you may not even understand, you pay ten dollars for one app that meets your basic needs. Apps are also easy to learn, since developers often distill the essence of application to the bare minimum. While websites often offer a bewildering array of functionality and corresponding complexity, apps deliver far fewer functions, but ones whose purposes are often targeted to perfection.

For instance, one Sunday a marketing executive found himself working in the office. He booted up his computer, glanced at his calendar, and realized to his chagrin that it was Mother's Day. Restaurant reservations go fast on this holiday, so he faced a difficult challenge. Although he was equipped at his desk with a dual core Xeon-equipped PC, a 21-inch high-definition monitor, and 45M of Internet bandwidth, he turned to his iPhone.

It was far easier to use the OpenTable mobile app than to search the web, visit restaurant websites, read reviews, select a restaurant, and then make the reservation. With the app, in less than a minute, he found an Italian restaurant (his wife likes Italian), that looked cozy (pictures included), located in McLean, Virginia (four miles from where he was sitting), with good food (reviews included), seating availability at 7 p.m. (when his wife prefers), and with a cost rating of $$$ (because he loves his wife).

After he tapped the "Make Reservation" button, the app asked if he wanted flowers delivered to his table—which he did. The app did one thing remarkably simply and fast, and then offered a bonus. What do you think this implies for the future of websites? I think every company that serves consumers will need a great app that offers their core services on their customers' mobile phones.

If they don't, their competitors will.

The App Store: Apple spearheaded a new model for software purchasing with the introduction of its App Store in July 2008 coincident with the introduction of its iPhone 3GS that came preloaded with App Store functionality. This micro-transaction system enables easy, one-stop application purchasing. The customer has a single credit card on file, one password, and there's only one site interface to learn for all applications. There's no need to hopscotch the Internet to make purchases. It's like a department store, minus the walking. Apple also claims to impose quality control on all merchandise, shielding buyers from malware. By October 2011 people had downloaded 18 *billion* apps from the App Store, or two and a half apps for every person on earth.[21]

The easy, one-stop buying concept behind Apple's App Store has served as the model for other application stores. Google's version of the App Store, the Android Market, reached 10 *billion* application downloads in December 2011. And phone companies like Verizon, Nokia, Windows, and Samsung have their own app stores as well.

Sensing the World Nearby: Thanks to the GPS, mobile devices know their location. Using gyroscopes and accelerometers, they know their orientation. They can hear voice commands through their microphones.

They can see with their camera eyes and interpret scenes, products, and bar codes.

As a result a phone can read the world—and more. You can point the mobile device at the stars, and it will show you a map of each constellation. Since the gyroscope knows the slant of your phone and the GPS knows its location, the device can match the image to an astronomical database and bring up the appropriate map. Then you can zoom and pan. Want to know the name of that glowing star to the south? Touch it and you'll see an answer.

And once it reads the world, the smartphone can *change* it for you, with types of software that can be called "alternate reality." For instance, if you were in Mexico City and gazing at a sign in Spanish, utilizing a program called "Word Lens," you could point your mobile device at this text. It would blink twice, and suddenly you'd be reading it in English.

These are all breakthroughs in human-computer interaction, and they're being propelled by mobile computing. They will change everything from healthcare to retail to the reading material you hold in your hand.

CHAPTER 3

PAPER

The Demise of Paper

Paper Vanishing Today – From Clay to Paper – Gutenberg and the Second-Copy Revolution – The Information Era – The First Post-Gutenberg Breakthrough: Desktop Publishing – The New Map – The New Book: From Tablet to Magic Tablet – The New Book Store – The New Library: Everywhere and Nowhere – The World as Your Library – The New Newspaper: All the News That's Fit to Post – The New Magazine: A Golden Age of Niche Publications? – The Paperless Office – Saving the Air

"I can't speculate on the future," Seth Kursman, a vice-president at AbitibiBowater said as the company filed for bankruptcy in April 2009. "Certainly we'll be putting together a plan to restructure the company and make it stronger."[1]

AbitibiBowater was no ordinary company. It was a giant—the largest maker of newsprint in North America, the supplier for 43 percent of the market in the United States and Canada. When it emerged from reorganization in December 2010, its chairman stressed a need to diversify.

But that won't be enough—unless it diversifies into something like mobile software.

For paper is vanishing into the world of papyrus and parchment. For the first time since the Crusades, there is a better platform for the written word. Our society prints billions of pages every year, and we're headed to zero. Books are becoming software. Newspapers are becoming software. Magazines are becoming software. Marketing brochures are becoming software. Maps are largely software already, and few would dream of using a physical map if they had a Google map.

A printed page is just one static page. A mobile screen can be every page ever written, delivered anywhere at any time, equipped with myriad interactive powers.

The death of paper may seem like a wild idea. We live in a world of paper. It's at the doorstep in the morning when the newspaper arrives. It's in the magazine rack at the checkout line. It surrounds us at Barnes & Noble. It's in envelopes in the mailbox, notebooks at school, and reports at the office. Indeed, global paper use grew more than sixfold in the second half of the twentieth century. In 1997, the world consumed some 300 million tons of office paper alone,[2] an amount that could fill 383 Empire State Buildings or make a pile that reached to the moon and back eight times. Each year, we publish more than two billion books and 350 million magazines.

Yet paper's downward slide was already underway before the mobile computing phenomenon arrived. For instance, sales of commercial printers dropped more than 11 percent from 2000 to 2010.[3] Do you still send Christmas cards? One out of every five people who sent them in 2005 had halted by 2011, as email and—especially—social networking made them unnecessary.[4]

But the real assassin is the e-reader. As long as the screen caused eye strain, paper had a chance. With the emergence of the Kindle and e-ink, the screen became pleasant to the eye. Suddenly, after a fabulous career as the carrier of content, paper became a burden.

And paper has many drawbacks. It's a static holder of ink patterns, so you can't push a button on paper. You can't search. You can't drill down for a definition or copy a paragraph. You can't animate paper. And it's expensive and noxious to produce. You have to pay people to chainsaw trees, and to manufacture the paper, ship it, print on it, package it, and deliver it. Then you may have to pay a retailer to sell it to you. Paper costs far more than a screen image, and it's far less useful.

Almost unnoticed, the Bureau of Labor Statistics has quantified the fate of the paper industry. Employment there, it says, will drop by about a quarter from 2008 to 2018, from 445,800 people to 337,500.[5]

The earth will benefit, for paper mills foul the planet. After the chemical and steel industries, papermaking is the third largest user of fossil fuels in the developed world.[6] The manufacture of books and newspapers releases more than 40 million metric tons of carbon dioxide annually in the United States, as much as 7.3 million cars,[7] and paper mills spew a toxic mix of chemicals into the air and water.[8] Despite everyone's interest in preserving green space, pulp mills have consumed some 5,000 square miles of U.S. forest every year, all for paper.[9]

From Clay to Paper

If you've ever drawn a diagram in the dirt, you know how the first print medium arose. When the Sumerians invented writing around 3000 B.C.,[10] they used the earth around them. They created wet clay tablets, pressed wedge-like characters called cuneiform into them, and baked them in an oven.

By the third millennium B.C. an improvement had appeared in Egypt: papyrus. This proto-paper derived from a reed (*Cyperus papyrus*) that grew in the Nile. The Egyptians cut strips from it, laid them in slightly overlapping rows, and hammered them to release juices that formed a glue that held the strips together. Papyrus was cheap and lightweight, but fragile. It could last in the arid climate of Egypt, and papyrus documents filled the library of Alexandria. But in damper climates such as in Rome, it quickly degraded and gave scribes much employment in recopying.[11] Papyrus was also brittle and would prove utterly unsuited to the rising book format.

According to one account, the Ptolemies of ancient Egypt took such pride in Alexandria's library that they tried to stop the rise of

FIGURE 3.1
Papyrus was extremely brittle and sensitive to its environment. Dry environments were best, as papryus did not last long in damp places like Rome.

similar archives in Rhodes and Pergamum. Since they controlled the papyrus market, the Ptolemies tried to lock out their archival competitors by forbidding its export. Ultimately, though, this tactic ended their monopoly, for Pergamum responded by inventing a substitute: *charta pergamenum*, or parchment.

Parchment is dried, de-haired, polished sheepskin, while its cousin, vellum, comes from lamb and calf hide. Both are flexible and hardy—some parchment documents from the fourth century seem barely a day old.[12] Its excellence stems from the fibers elastin and collagen. Elastin makes skin flexible enough that it bounces back after pinching, and collagen, one of nature's strongest proteins, gives strength and durability. Contracts on parchment date to 258 B.C. The Dead Sea Scrolls are parchment, and by the fourth century A.D. parchment had ousted papyrus in Europe.[13]

However, both parchment and vellum were expensive. To print the equivalent of the Gutenberg Bible on vellum, it would mean having to slaughter 200 animals. But the meadows of Europe could support only so many sheep and calves, so supply was capped.[14]

Meanwhile, Asians had long been using a better substance. In 1957, archeologists discovered a four-inch-square of paper in a tomb near the ancient Chinese capital of Sian. Dated between 140 and 87 B.C.,[15] it's the oldest surviving piece of paper.

In fact, the Chinese had invented paper earlier, around 200 B.C. These first sheets were thick and uneven, so people used paper mainly as a wrapper. They wrote on silk or bamboo instead, and the bamboo strips holding the imperial history of Sima Qian (ca 145 or 135 B.C.– 86 B.C.) could fill a cart.[16] In contrast, the coarseness of early paper suited it for clothing and, by the ninth century, even armor, since certain pleated kinds of paper actually repelled arrows.[17]

Paper traveled west slowly, the secrets of its manufacture guarded closely. It reached the Islamic world in 751, and by 900 Baghdad had public libraries fed by more than 100 book manufacturing shops.[18] In 1009, the Moors established Europe's first paper mill at Xativa, near Valencia in Spain.[19] By the twelfth century, the technology reached northern Europe.

There it entered a world whose information poverty we can hardly imagine. Books were rare and precious, partly because their creation was so labor-intensive. Monks hand-copied documents from dawn to dusk, working in silent, cloistered rooms called scriptoriums, and a single, richly illumined book could take years to create.[20]

Paper made quick inroads. It was less durable than parchment, though more flexible, but its overpowering advantage was price. Its source—rags, at first—seemed inexhaustible. Parchment retreated into specialty uses such as diplomas, government documents, and the surfaces of drums and tambourines, while paper quickened European commerce. Economies run on information, and paper made it cheaper and faster for information to spread. The demand for scribes increased, and many moved out of the monasteries and into towns, where they also fed the economy by spending money. The number of books grew, universities took root, and international trade in the fourteenth and fifteenth centuries rose significantly both in value and volume.

Even so, until the midfifteenth century, virtually all books in Europe were still handwritten and costly.

Gutenberg and the Second-Copy Revolution

Printing wasn't new. The Chinese invented it sometime between 704 and 751,[21] and the first known printed book, the *Diamond Sutra*, appeared in 868. Soon after, biographies, literature, and philosophical works tumbled forth, as well as calendars, playing cards, and paper money. In those days, printers carved whole lines or pages of text into woodblocks and might run off a million copies or more. But, the general rule was print-on-demand. Since each inscribed woodblock remained whole, printers could set them up as quickly as was needed, and families might hold onto them for centuries.

The Chinese also invented movable type. Bi Sheng, a man virtually unknown in the West, developed it between 1041 and 1048,[22] but the technique spread slowly, hampered by the complexity of the Chinese language itself. The Chinese script uses logograms, characters that usually represent a spoken syllable attached to a meaning, and it has some 47,000 of them. Moreover, a printer couldn't have just one set of 47,000 character dies, since a word like "love" might occur several times on a page, so printers might stock twenty different dies for each such logogram.

Thus, for instance, in 1725 the imperial printing works kept 200,000 bronze dies on hand. In addition, many logograms were rare and known mainly to scholars, so Chinese printers needed special expertise. They couldn't be blue-collar workers, so costs were even higher.

The character sets in European languages were far simpler, and the man who would make information explode in the west is one of history's great mysteries, a figure even hazier than Shakespeare.

Johannes Gensfleisch zum Gutenberg was born between 1397 and 1403 in Mainz, on the Rhine River, and received a good education. He probably trained as a goldsmith and learned metalworking. Around 1448—the timeline is a guessing game—he began the experiments that led to movable type, which according to one authoritative source emerged between 1451 and 1455.[23]

Like "mobile technology," "movable type" is shorthand for several innovations. Gutenberg brought three technologies together: movable character dies, the right kind of ink, and the printing press. Startup and storage costs plunged with movable type, since it was reusable. And the press itself, which he adapted from a winepress, was like a rubber stamp on a hinge. Printers could reproduce an identical page in moments, and make myriad copies at little cost.

So the great advantage of the printing press lay not in the first book, but in the second, and third, and all the rest. Marginal production costs dropped tremendously. For instance, in 1483, the Ripoli Press charged three florins per quinterno—that is, five sheets of paper folded in half—to print a single copy of a translation of Plato's *Dialogues*. A scribe would have charged one florin to write the same amount, so the handwritten version was actually cheaper. But, what about the next copy? The scribe would employ the same labor and charge the same price: one florin. The printer, on the other hand, could slash the price on that second copy and, as it happened, Ripoli produced 1,025 copies of Plato's work.[24]

When a best-selling author spurred demand for large numbers of books, the price got *really* affordable. A handwritten Bible cost between 60 and 100 guilders in Germany in 1422, but Martin Luther's popular translation of the New Testament cost only one and a half guilders in the next century.[25]

The Information Era

Information is subtle and easy to overlook, yet fundamental to everything. It is the ghost in the machine of markets, economies, and societies. Gutenberg's innovations that spurred a new era, prefigures some of what we can expect from mobile technology.

First, like mobile technology, printing finally made information portable. As more and more titles poured off the presses, books expanded beyond the confines of the library and access to information soared. Before Gutenberg, some 30,000 books existed in all of Europe,[26] but by the year 1500, European printing presses had churned out between 10 and 20 million volumes.[27] In the following century, European output rose to between 150 and 200 million volumes.[28] In the span of just 150 years, the number of books grew at least 5,000-fold, and by 1600 there were two books for every person in Europe.[29]

Printing also caused an explosion in book variety. By 1600, European printers had pumped out 1.25 million *titles*,[30] a spectacular increase compared to those initial 30,000 pre-Gutenberg manuscripts.

Printing brought information to the public *faster*. The scriptorium was a time-sink. For instance, the library at Cambridge University had 122 books in 1424, and over the next 50 years increased that number only to 330[31] thus producing only four books per year on average. Early printing presses required about an hour to set up a page and could then produce up to 240 prints per hour, which equates to a production of six books of 200 pages each per day.

The spread of information led to the Renaissance and its exhilarating sense of discovery, to astronomers Tycho Brahe and Johannes Kepler, to Andreas Vesalius and his modernization of anatomy, to William Gilbert and his scientific approach to magnets, and to William

Harvey and his discovery of the circulation of blood. It led to Galileo, who would publish one of the first popular science works, *Dialogue Concerning the Two Chief World Systems*. Such advances lie at the base of our understanding of the world today, and they could not have been birthed from the scriptorium.

All of these improvements made people wealthier. By one analysis, in the late 1400s cities with printing presses grew at least 60 percent faster in the following century than similar cities without presses. From 1500 to 1800, such cities grew at least 25 percent faster, as printing endowed residents with more information, so more of them developed skills important to commerce. They more quickly and continually learned facts that could lead to wealth, and they created innovations that further spurred trade and prosperity.[32]

Johannes Gutenberg himself did not grow prosperous. After he printed his showpiece Bibles, his business partner sued him and seized the print shop. Gutenberg seems to have continued printing on a small scale until 1460, when some think he went blind, and he died in 1468. He must have known he had changed the world, but he could hardly have guessed how much.

The First Post-Gutenberg Breakthrough: Desktop Publishing

Though the copier made everyone a micro-publisher, the next real printing revolution occurred with desktop publishing. Its first incarnation was the dot matrix printer, then the laser printer improved it significantly, and color printers made it better still.

Desktop publishing gave everyone true typographic control, and more important, it took the user from uniformity to mass customization. The printing press let an individual print 1,000 copies, all the same. Desktop publishing let a thousand individuals print one copy, each *different*. It could also let an organization print a thousand variations of a single report, with content tailored to a thousand different recipients.

Yet desktop publishing didn't change the *cost* of distributing information on a broad scale. Today, a new hardback book costs the consumer $26 or so, and even if that title might be downloaded for free, by the time you print it out on paper from your desktop printer, the cost of the paper, the printer, and the ink cartridge would likely be $26. With desktop printers, ink alone is more expensive than blood.[33] Hence, print-on-demand books usually cost more than books mass-produced from an offset press.

Nevertheless, with desktop publishing everyone had a press, so everyone became a printer, and the use of paper soared. Indeed, by 2010 the average office worker in the United States was generating about two pounds of paper and paperboard products each day.[34] That amount would have been inconceivable in the days of the typewriter.

Now, with the emergence of app-phones and tablets, the desktop printer is about to go the way of the typewriter. With maps, the reason becomes obvious.

The New Map

In 227 B.C., Crown Prince Tan of Yan was required to endure a humiliating submission to the ruler of the Qin state. The first step was presenting a gift to the hated potentate, and the prince did not send an ordinary man for the task. He sent the learned Jing Ke—as an assassin. The plan was simple. Jing Ke hid a dagger in a case, along with a district map. As he presented the Qin ruler with the map, he would slip the dagger out and slash vital organs. As it turned out, the killer's aim went wild, and the scheme collapsed—but it left us with the first Chinese mention of a map.

Of course, maps are far older than this plot. The earliest known ones adorn the caves at Lascaux in France, date from 16,500 B.C., and show stars in the sky. The Mesopotamians made maps portable. They drew them on clay tablets about the size of a hand, and one from 2500 B.C. depicts a river between two hills, with features such as a stretch of land owned by a person known as Azala. The ancient Egyptians and Greeks used papyrus, of course, and one of the oldest pieces of paper from China has a map on it.

The paper map was a fine product once, but it's a ridiculous one today. Most people can't even fold one correctly. Each map has a single, fixed scale, so it's almost always wrong: either too high with a paucity of detail or too low with a swarm of it. It's out of date by the time it is printed. It's an extra object to carry around, it becomes clutter in the glove compartment, and it costs several dollars.

But a Google map costs nothing. It adds no weight and takes up no space. It can be a roadway schematic or an aerial photo, or even a panoramic terrestrial photo. You choose its scale by moving your fingers, so you can zoom down to a single city block or up to view an entire continent. It can have unlimited, contextual advertising, and can tell you where the nearest restaurant is, how to get there, and how much time you'll need for transit. It even can substitute for travel itself, since it shows the facades of streets throughout the world. It can talk to you. It can listen to you.

It's an *intelligent* map. Any way you look at it, it's better than the paper item it's replacing.

Every mobile document can have a similar impact.

The New Book: From Tablet to Magic Tablet

What is a book?

The easy answer is that it's an object with pages of text, and perhaps pictures, all bound firmly at the edge. But that's the delivery system, or the container of information. There have been three such containers.

The Mesopotamians put books such as *The Epic of Gilgamesh* on sets of clay tablets about three inches across. They could be carried, but they were weighty, each was discrete, and they needed baking (or at least drying). They had two big advantages. They were permanent, and they improved on the alternative: no container at all.

Tablets didn't disappear with the Mesopotamians. This format excelled for short documents, and the ancient Greeks and Romans sent letters on small, erasable beeswax tablets.[35] Pupils in nineteenth century U.S. schools used erasable slate tablets, now sold as antiques.

Scrolls are another container of book information. They arose with papyrus in Egypt, and we now identify them with white-bearded sages from antiquity. The text appeared in lines parallel to the edges and stacked in columns. The unread part was held in the right hand, and the read portion in the left. Thus the reader slowly unwound the eighteen-to-twenty-five-foot document. It was like manually turning the tape in a video. When done, it was rolled back up and stuck with other scrolls in a box. Plato's dialogues originally appeared on such scrolls, and in one of them Socrates made a declaration.

"Nothing worth serious attention has ever been written in prose or verse."[36]

Even so, the scroll was a big improvement over clay. It united the entire work in one object. It was a less expensive delivery system, and much lighter and easier to move around.

And it is a fossil.

Today's book format is the codex. It arose in the Coptic monasteries of Egypt shortly before the birth of Christ, and by the third and fourth centuries it had swept across the European world. Instead of rolling the material on a scroll, its page-like parts were cut out and glued together at the spine. The only mystery about the codex format is why it didn't appear earlier, for it had every advantage over its predecessors.

The scroll was clumsy; the codex was a snap. The scroll invited ruinous damage, since a rip could tear it in half, while the codex limited most harm to a page. The scroll allowed writing on just one side; the codex allowed it on both sides, so it was more compact. The scroll limited book length; the codex extended it. The scroll gave you a sequential view of the text; the codex allowed browsing and flipping through the content. It also organized the text better, since it allowed

FIGURE 3.2 Scrolls were a big improvement over clay. The scroll above is an Amharic manuscript from Ethiopia.

page numbers, an index, and a table of contents, all of which sped up information access.

By the third century, the codex became crucial to the spread of Christianity, partly because a forbidden codex was easier to hide from Roman authorities.[37] For all these reasons, the codex prevailed for two millennia. It has had so few rivals that the term "codex" fell into disuse, and most people don't even know what it means.

The PC screen arrived in the late twentieth century. It had many advantages over the codex, but certain basic drawbacks. No one ever took a desktop PC to the easy chair to read, and even a laptop was bulkier than a book. There were no applications designed specifically for reading books, but the key problem was eyestrain. Pixels on a computer screen are backlit, illuminated from behind, so you are staring at glare. A big step forward came with e-ink, first developed by Joseph Jacobson at Stanford in the 1990s. E-ink forms characters out of tiny bits of dark material the computer shifts about. It works like the cards that fans in a stadium hold up to create large images. E-ink is reflective, so it's easier on the eyes, and it works even better in bright light.

It has brought us from the clay tablet to the silicon tablet.

Now the mobile screen makes the codex format as obsolete as the scroll. It boasts many advantages:

- The content can be organized any way the author wants. The software can imitate a codex, a scroll, or even a clay tablet.
- It can support a linear path through a book like a codex or scroll, or it can provide hypertexting that allows a multitude of paths through the content.
- The reader can look up words and dig deeper for information in an instant.
- There's no need for costs of a printing press, paper, ink, or electricity, lubricants, cleaners, and labor to rub the printing process.
- There's no cost for the shipment of books. Vans and drivers disappear.
- The cost for storage disappears. For the publisher, warehouse space is no longer a concern.
- There's no need—or cost—for people to sell the books. Salary, inventory, and real estate costs plunge as bookstores become artifacts of the past.
- The purchasing process is faster and easier. There's no need to drive to the bookstore and wait in line, or wait for an online retailer to mail a bound book. A book can be purchased anywhere and at any time.

In other words, e-books on mobile tablets change everything. To an e-publisher, a building is no longer an asset. Typeset is not an asset. A printing press is not an asset. Inventory is not an asset. Less capital is needed. Certain legal skills remain an asset—such as copyright knowledge and negotiating skills—but everything involved with the manu-facturing of books has become a liability.

How much do e-books save? Figure 3.3 shows cost breakdowns in 2011. The big dive comes with the bookseller—the owner of your local bookstore. If you pay $26 for a hardback in the bookstore, the book-seller gets $13. Buy online, and the bookseller receives $3–4. The other figures are roughly equivalent, though the author receives less and ex-penses drop for the cheaper e-books.

Publishers earn slightly more with e-books, and they also gain secu-rity. Before e-books, with every hardback they shipped to a store, the publisher paid the cost of a print run and the author's advance.[38] Yet the bookstore could return that unsold book for a full refund. Most books lose money, but best-sellers are jackpots that often cover those losses.

With e-books, the print run disappears. Returns disappear. Pub-lishers can risk less, yet enjoy the profits they make from the sales.

The math is so overwhelming that some believe, in the end, book-stores will live on like black rhinos in a wild animal park, by the grace of outside aid. Some see this as the reason behind the artificially high price of some e-books.

"If you want bookstores to stay alive, then you want to slow down this movement to e-books," said Mike Shatzkin, a consultant to pub-lishers. "The simplest way to slow down e-books is not to make them too cheap."[39]

But attempting to slow down digital book sales is futile. By Octo-ber 2010, digital books comprised 9 percent of the total book mar-ket[40] and was growing vigorously.

The New Book Store

As bookstores and other middlemen in the publishing process dwindle away, where will the money be made?

It's easier to predict the future than to make money from it. Some think the best bet is hardware—the e-reader—though the e-reader

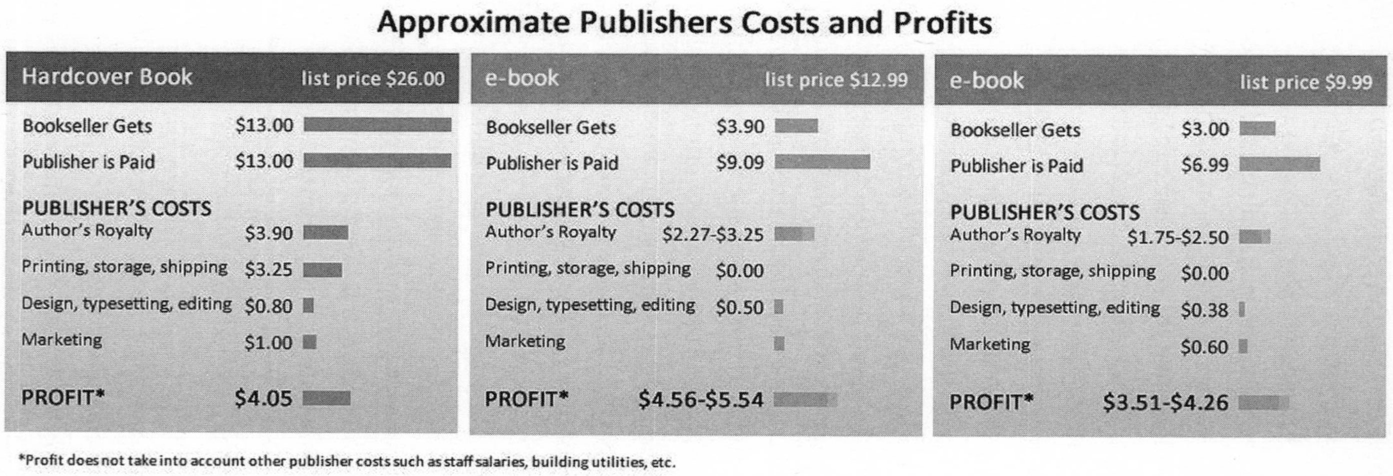

Approximate Publishers Costs and Profits

Hardcover Book	list price $26.00	e-book	list price $12.99	e-book	list price $9.99
Bookseller Gets	$13.00	Bookseller Gets	$3.90	Bookseller Gets	$3.00
Publisher is Paid	$13.00	Publisher is Paid	$9.09	Publisher is Paid	$6.99
PUBLISHER'S COSTS		**PUBLISHER'S COSTS**		**PUBLISHER'S COSTS**	
Author's Royalty	$3.90	Author's Royalty	$2.27–$3.25	Author's Royalty	$1.75–$2.50
Printing, storage, shipping	$3.25	Printing, storage, shipping	$0.00	Printing, storage, shipping	$0.00
Design, typesetting, editing	$0.80	Design, typesetting, editing	$0.50	Design, typesetting, editing	$0.38
Marketing	$1.00	Marketing		Marketing	$0.60
PROFIT*	$4.05	**PROFIT***	$4.56–$5.54	**PROFIT***	$3.51–$4.26

*Profit does not take into account other publisher costs such as staff salaries, building utilities, etc.

FIGURE 3.3 Publishers can increase profit with a switch to e-books.

Data Source: New York Times, see Rich.

tablet that delivers only books is likely a dead end. Why buy both an e-reader tablet and a tablet computer if the computer can be a book as well as innumerable other things? Why carry both? The tablet computer will absorb the e-reader, and the term itself will become a nostalgia item.

Some think it's an e-bookstore. But Apple has one, and it could eventually surpass Amazon's, even though Amazon beat Apple to the market by a decade. There may be room for one, two, or even three e-bookstores, but not many more. Look at music. Everyone knows Apple iTunes, and Amazon is number two, but can you name the third biggest online music store? There's no enthusiasm for it, and it's highly unlikely we'll see a fourth, fifth, or sixth.

The e-bookstore model itself could face challenges. Content aggregators will arise—book versions of MetaCritic—summing up book reviews and steering traffic directly to author or publisher sites. And other individuals could flex market muscle. Well-known personages might realize that millions of fans would love to know what they're reading, and would buy the same books. Mini versions of the Oprah Book Club could spread. So, an array of individuals could take on the bookseller role in recommending books and directing consumer demand.

What about publishing? In the days after Gutenberg, publishers were also booksellers. Today, publishers could sell e-books directly, without shouldering the cost of middlemen. But they face two obstacles: traffic and price pressure. Amazon sells *all* titles, and that generates traffic. It's more convenient than a publisher's site, which would sell just those the publisher owns. On the other hand, customers could hijack retail from Amazon in the same way that Amazon hijacks retail from Best Buy today. They could browse the Amazon site to identify a book they want, then buy it less expensively directly from publishers, much as they have browsed Barnes & Noble's physical bookstores, and then bought more cheaply from Amazon.

The price of e-books is another issue. It could continue to drop until it approaches that of songs, around $1.99. Public domain and

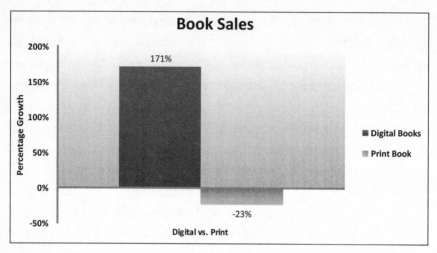

FIGURE 3.4 The graph shows the percentage growth in book sales from October 2009 to October 2010.

Data Source: See Pham.

other out-of-copyright books are already free online, as they never would be essentially given away in the brick-and-mortar stores, where printing and inventory costs are still incurred by the retailer. Moreover, a lower price for e-books might encourage massively more consumption as we have seen with mobile apps. People will click on a $1.99 item without thinking too much. It feels almost free, so they'll give it a try. But $19.95 is a much more significant commitment to a piece of work one might not ultimately value. The fact that $1.99 works economically for mobile apps, implies that it should also work for books since applications cost a lot more to develop than books.

For example, you can buy a word processor app from Apple for $9.95, and yet it costs much more to produce than any book. Books normally require one to three man-years of labor, whereas most good software programs require five, ten, thirty—even fifty or a hundred man-years of development. And so, for the book industry to survive, the prices may need to plummet, and they probably will as the economies of scale increase and the market for e-books grows.

Publishing may face a squeeze.

And what about the author? Well-known writers already are going straight to their readers, and others could do so, too. We're seeing the trend of the micro-celebrity—little-known authors who generate a great deal of publicity (and sales). If you are a professor at a university, you have a captive readership: your students. So you could set up a Facebook page and post your books there. You could self-publish through Apple or Amazon, direct traffic to those sites, and collect 70 percent of the income. Or sell directly on your website and collect 100 percent.

Suppose you're Dennis Conner, the winner of the America's Cup. You could sponsor a whole *series* of books on yachts, boat building, and sailing, as well as conducting lectures, and producing videos and movies. If you're a film star with an autobiography, a sound track, and three movies, there's no reason you couldn't upload all of them to the iTunes and the iBook stores, market them yourself, and pocket a hefty royalty. Though you wouldn't own the copyright to the movie, you could cut a deal with the studio and keep a percentage of the profits, as Amazon does.

Any business whose assets consist of relationships with authors, marketing, and editorial know-how will probably continue as a going concern. There's room for a middleman who can provide editorial or marketing guidance, but middlemen who provide manufacturing, distribution, or retailing expertise will be driven out. Apple and other e-bookstore venues will probably take over that job, and perhaps Amazon as an e-bookstore—but not the Amazon that mails out physical books.

In the end, nothing will remain but writing, editing, perhaps some design, and marketing. Pure information.

We are a long way from clay.

The New Library: Everywhere and Nowhere

Aristotle's most famous pupil was Alexander the Great, who seemed to show little fondness for the scroll. Yet Aristotle would exert vast influence through him. After the fiery conqueror died at 33 in 323 B.C., three generals carved up his empire. Ptolemy I Soter got Egypt and exalted his ties to Alexander by creating the first world library, for the two had been childhood friends and he too had studied with Aristotle.

The fabled institution gathered some 700,000 scrolls over the years, concentrating huge amounts of human knowledge. It attracted great minds like Euclid, Galen, and Archimedes and became the model for the vast university research libraries of today.[41] Yet, it hardly held all written knowledge, and the state guarded its scrolls closely. It was never a true world library.

Today, however, we are on the brink of one. As physical libraries close, we are seeing an explosion of accessible information dwarfing anything that came before.

The law library foreshadowed this trend. Attorneys once went to physical library locations to do research, discuss legal issues, and socialize. It was the "public square" of every big law firm. But the research took time, and lawyers were willing to pay handsomely to have a faster process. So online services like Lexis and Westlaw appeared, and as the cost of print materials and office space rose, law libraries became far less necessary. In-house libraries started shrinking around the early 1990s.

"Today law librarians are often more concerned with Wi-Fi reception than full sets of [case law reporters]," an observer noted.[42] Meanwhile, lawyers have far more information at their command than ever before, available online in a flash.

Traditional libraries now stand where law libraries did in the early 1990s. At the massive Central Library in downtown Los Angeles, fewer browsers stroll the floor, and most reading tables are empty. In some university libraries, reading rooms sit almost vacant as students conduct research from their rooms. And, patrons are borrowing fewer books. During the 2009–2010 fiscal year, checkouts fell by a million volumes at the New York Public Library, and checkouts at the typical library declined almost 6 percent between 1997 and 2007.[43]

Libraries are attempting to respond. They are loaning e-books that "self-destruct" after a few weeks. Though some 98 percent of library shelves remain stocked solely with physical books, from 2005 to 2008 the number of paper books being added to U.S. libraries flatlined. At the same time, their e-book collections grew almost 60 percent.[44] Libraries are also broadening their mission, turning themselves into computer centers. The L.A. Central Library has an often-crowded seventy-seat area where people scour the Internet, check Facebook pages, and watch YouTube clips. Some libraries have created game rooms, computer clusters, and Internet cafes. They've also increased their number of DVDs and high-definition TVs.

But their most important response thus far is digitization of their catalogs.

Libraries are offering e-books for loan from their websites, creating electronic caches you can reach from your home. The process is expensive, and it is proceeding slowly, but channels are opening that will grant access to the rare and wonderful. At Harvard, for instance, specialists have digitized the fragile pages of *Story of Red Plum Blossom*, a 400-year-old Chinese drama known mainly to researchers. Eventually the university plans to scan 51,500 rare Chinese books in the Harvard-Yenching Library, perhaps the largest collection outside of Asia.

The idea of digitizing books is at least forty years old. In 1971, Michael S. Hart was studying at the University of Illinois, where he got access to a mainframe. He realized that many ordinary citizens would soon own computers, and he decided to make public domain books—those no longer protected by copyright—available for free. He called the effort Project Gutenberg, and began by digitizing the Declaration of Independence. Today, the Project has more than 34,000 books that anyone can download at no cost whatsoever.[45]

In 2004, Google launched a much broader digitization effort. It started scanning books, working with libraries at universities such as Harvard, Stanford, Oxford, and Cornell, and thus far the collection boasts more than 15 million titles in more than 300 languages.[46] The company plans to offer this database to researchers, those wanting to examine developments such as changes in language and testing literary theories by analyzing text. Already, on its Ngram Viewer, anyone can type in a particular word and see how its usage has evolved over the last few centuries.

The entire Library of Congress could be made accessible by your mobile device. It's the world's largest library, with 530 miles of shelves holding more than 100 million items: books, journals, periodicals, maps, art, and designs—information of every sort. Assuming that the average price of an item from the collection is $10, then digitizing this library would make available a $1 billion resource. For the total world-wide population, that's material amounting to $7 quintillion ($1 billion times the 7 billion people on earth). Of course, the sum doesn't take into account all of the secondary benefits of knowledge.

Networked devices will eventually be able to fetch every surviving page written and produced in every language since the beginning of history.

At the same time, we're seeing the emergence of world libraries of other kinds. Wikipedia is building a world encyclopedia that grows bigger every day, and eBay is creating a world catalog of discounted goods. Apple has the iTunes University, which is becoming a world library of video lectures. Google's street view project is a world library of pictures of byways all over the earth, while Google Maps and Google Earth provide a world library of maps and satellite images.

The result will be an immeasurable storehouse of knowledge, available on a device slightly larger than a library card.

Physical libraries will become obsolete. They won't turn into computer centers, because everyone will carry their computers with them. The lending of e-books will cease. Many books are in the public domain and free, and as for the rest, when e-books cost just a few dollars, most people will simply buy them. In its very death, the library will realize its most spectacular success.

The World as Your Library

Mobile technology presents the opportunity for a new type of library, a true "world library" that Ptolemy could never have accomplished in the physical world. Using mobile technology, simple bar codes or tags based on Near Field Communication (NFC) technology can direct mobile devices to access information much more swiftly than any card catalog.

As information becomes vapor, it will surround us and we will breathe it in everything we do.

Take the Finnish city of Oulu, for instance. Oulu lies 125 miles from the Arctic Circle and has about 140,000 residents. It's been one of Europe's "living labs," where whole communities experiment with new technologies. In 2006, they installed 1,500 tags in buses and restaurants, at bus stops, and in one pub and theater. Citizens used NFC to pick up bus schedules and maps. They waved their mobile phones at a poster of a theatrical play to learn more about it. In schools, students touched tags to get information about schedules, coursework, cafeteria menus, and daily announcements.[47] Technologies like NFC make it easier to access information about almost everything, and eventually the tendency will be to electro-tag the world.

If we can electro-tag the living world, why not the past, as well? After most people die and a few years pass, the world knows little about them except what their headstone looks like in the cemetery. But what if tombs could talk? What if people could walk up to your headstone, see you, and get to know you, long after you had died?

In fact, you already can. You can have your grave tagged so anyone can walk up to it and link it to a "digital tombstone," using NFC, RFID, a barcode, a GPS reading, or even just a searchable name. The visitor might see a video epitaph, an autobiography, a

FIGURE 3.5 Scan this apple with your smartphone and learn the cost, weight, origins, and shelf age of the apple along with some suggested apple recipes.

message to loved ones, and a reflection on the nature of life and death. Every person could prepare a record, and the images could last forever, bestowing a kind of immortality. The concept has something romantic about it, especially the idea of leaving a memorial for your children, grandchildren, and posterity.

Professionals might help people create these films, much as videographers shoot weddings and other special events.

And with video epitaphs, no one will even have to travel to the cemetery. In 2010, 32 percent of Americans who died and more than 70 percent of the British chose cremation.[48] For all of those people, their headstones could be replaced by their Facebook profiles, where friends and family post memories and notes of condolence. Anything they can put on your tombstone can be put on the wall, and much more—videos and photos, your favorite films and books, your responses to people and events. It could become a time capsule, where a figure—famous or otherwise—records a tale to be unlocked years after death.

"Here's my complete story, the facts I couldn't tell while others were alive." There's no reason your epitaph has to exist physically, when it can live in cyberspace forever.

This idea need not be limited to the deceased. You could have a compelling professor of Roman history walk with you through the ruins of Pompeii. The paintings there, on the walls at the Estate of Julia Felix, are now so faded they are almost blank; a typical tour guide might ignore them. But not an expert. With a wave of your mobile phone, the expert would tell you that an artist made engravings of those paintings in the nineteenth century. Then he would show them to you. Throughout the town, you'd see the ancient streets come alive with traders, hucksters, beggars, and ironmongers.

So why not get the story right, upload it, and let people reuse it a million times? Here, too, information providers will compete, and we'll see the demise of the mediocre. We can mass-manufacture a good tour of Pompeii.

Cyberspace is wrapping around physical space. In the past, everything on Wikipedia was trapped in your computer and before that, in a book at the library. Thanks to mobile technology, Wikipedia can leap into the real world and give you answers to every question you have and information about everything you see.

The New Newspaper: All the News That's Fit to Post

Printed books will linger on as luxuries, but print newspapers already are gazing into the maw of extinction. And they admit it.

"We will stop printing the *New York Times* sometime in the future, date TBD,"[49] its publisher Arthur Sulzberger has said. Some major dailies, such as the *Christian Science Monitor* and the Seattle *Post-Intelligencer*, have already gone all-digital. Denver's *Rocky Mountain News* shut its doors in 2009, and eight major newspaper chains including the Tribune Company declared bankruptcy between 2008 and early 2010.[50]

Part of this turmoil stemmed from the economic downturn, but that was just grease on the slide. The paper model of news is like the scriptorium in 1480. People are wondering how to scrap it.

Mobile technology excels in this area, because news breaks all the time—not just when the paper is printed on when you're seated at your computer. By 2010, mobile devices had already become the "most important medium" for getting news, according to a survey of 300,000 mobile customers. More than 30 percent of them ranked it first. The desktop PC was a close second at 29 percent, followed by television at 21 percent, and newspapers at a near-negligible 3 percent.[51]

For most of history, news has been oral, passed on by traders and simple word-of-mouth. Julius Caesar created a quasi-paper, the *Acta Diurna*, around 59 B.C. to notify the public about military campaigns, trials, executions, and similar events. Scribes wrote the news on white boards and placed them in crowded spots in cities. Since readers had to travel to see it, they bore most of the distribution costs.

Throughout the Middle Ages, the average European heard most of the news from sermons. But the printing press and the rising economy of Europe spurred a hunger for information. By the late 1400s, hand-written news sheets were circulating in German cities, and one offered lurid coverage of the deeds of Vlad Tsepes, now known as Count Dracula. In 1556, the Venetian government published *Notizie Scritte* ("*Written News*"), which readers bought for a coin or *gazetta* (pl. *gazette*).[52] Johann Carolus launched the first modern, printed newspaper, *Relation*, in Strasbourg in 1605.[53]

Early papers focused on highbrow content, such as financial news, but as the printing process grew more economical, expenses fell and the audience widened. With the so-called penny presses of the 1830s, newspapers spread through society. Vendors hawked these cheap papers in the streets, further widening distribution. In effect, the *Acta Diurna* was now coming to the audience. The telegraph enabled news to travel greater distances, at the speed of light, and made newspapers more valuable still, and the late nineteenth century saw the rise of news empires with names such as Hearst and Reuters.

Today, print papers survive purely on that momentum. Some 1,400 daily newspapers remain in the United States, as well as thousands of weeklies and biweeklies,[54] and their voyage into the void ahead illustrates how technology works to make the world operate more efficiently.

If news isn't new, it's history. Using a tablet device, I've often watched stories about stock market futures change before my eyes. While print news comes out just a few times a day, you can get mobile news wherever you are, often as it happens. In response, the newspaper has become more like a news magazine, with feature stories splashed across the front page. Since the important stories are stale by the time they're printed, papers lead with niche items that are new to the reader.

At the same time, print news costs far more than online news. For instance, a subscription to the *Los Angeles Times* costs about $350 per year in 2011, while the same paper is free online and has richer content. Google News is also free, and lets you read almost every major newspaper in every language on earth. So, the process of issuing print news is now largely a waste of time and money.

Here's how it works at the average big-city paper. First, journalists, editors, photographers, marketers, and layout artists create the content, the news stories and advertising. Blue-collar workers ink the press and run newsprint through it. Then they package the paper, organizing it into sections and folding it.

Next, truckers take the packaged newspapers to distribution centers. There, shippers re-bundle the packages according to home delivery routes and then haul them to drop-off points. The home delivery carriers arrive, load up their papers, and take them one-by-one to the doorsteps of subscribers.[55] Snarls can occur anywhere between production and final distribution.

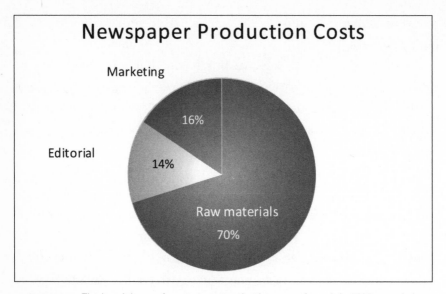

FIGURE 3.6 The breakdown of newspaper production costs from July 2009 reveals how raw materials overwhelmingly dominate the costs.

Data Source: See "High Operating Leverage Pressuring Newspaper Companies - Seeking Alpha."

This procedure takes a stunning toll. When the New York Times Company—a group of papers and other investments—released its quarterly report in September 26, 2010, it listed its cost for raw materials at $39,571,000.[56] In other words, if it had saved its outlays on paper and ink alone for two years, it could have bought a professional baseball team. Figure 3.6 shows the staggering weight of production costs on the industry, and this imbalance renders the newspaper industry paralyzed when trying to overcome competitive threats.

Meanwhile, as seen in Figure 3.7, ad revenue is evaporating and subscriptions are declining.

Ad revenue is evaporating. Ads have classically accounted for four out of every five newspaper dollars,[57] and in some cases the Sunday paper—with all of its extra sections and inserts—has brought in half of the revenue.[58] The classified section was once a major tributary to ad income, but from 2000 to 2010, almost three out of four want-ad dollars vanished.[59] The reasons are obvious. Craigslist is free for most

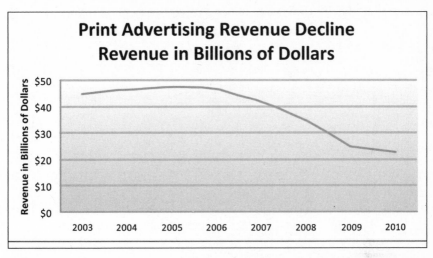

FIGURE 3.7 In 2000, annual ad income at newspapers peaked at $48.7 billion; by 2010 it was $22.8 billion.[60] This is a decrease of approximately 70 percent.

Data Source: Newspaper Association of America via Edmonds.

purposes, it's nearly instantaneous, easy to use, highly centralized, and widely viewed.

It's hard to think of any area where print has the advantage. So we're seeing obliteration here.

Subscription revenue is the other big part of newspaper income, and it, too, is diving. As shown in Figure 3.8, the newspaper industry lost 30 percent in daily circulation from 1990 to 2010.[61]

The falloff began in earnest in 2005. Before that, papers had been losing about 1 percent of their subscribers each year. The drop off pace increased to 2 percent in 2005, 3 percent in 2007, and 4 percent in 2008. In 2009–2010, weekday subscriptions at 635 papers fell an average of 8.7 percent. There was some relief during the six months ending in September 30, 2010, when they fell an average of just 5 percent.[62] Comparisons after 2010 become harder because the newspapers have relaxed the rules defining who is a "subscriber."

Newspapers responded to the assault with two desperate strategies: cut quality and charge more. First, they laid off journalists, trimming

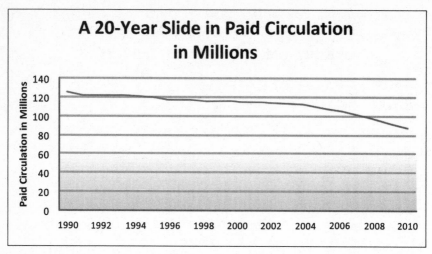

FIGURE 3.8 The newspaper industry lost 30 percent in daily circulation from 2009 to 2010.

Data Source: Editor and Publisher Yearbook via Edmonds.

newsrooms by 25 percent since 2000.[63] With fewer reporters, print papers are less informative and less valuable. Papers have also shrunk physically, and now have fewer and smaller pages. The Abitibi-Bowater bankruptcy stemmed from these reductions. Meanwhile, prices escalated, so more subscribers fled, revenue slid, and prices rose still higher.

It's a downward spiral into thin air.

The cost of distribution once segmented news markets. People in Dayton read just the *Dayton Daily News*. In Miami they read the *Miami Herald*, and in San Francisco and Los Angeles they read the *Chronicle* and the *Times*. Readers bought local papers because they had no choice. If you couldn't get the best, you took the local best, which was almost always mediocre.

But the newspaper business is not a manufacturing business any more. It's an online software business, and with software, local segments are not necessary. In the world of ubiquitous mobile devices, news distribution is instant, global, and almost free. You can get the best from anywhere, and local mediocrity is doomed.

Local news businesses can only survive by focusing tightly on a niche where it is in fact the best. The Portland *Oregonian* can be the best newspaper for local news, but it won't ever be the top international newspaper. The *Dayton Daily News* may survive to report on local cultural events and the sports scores of Dayton teams, and there is a place for that.

The same is true for national and international news. For political news, for example, the average business executive will read the *Washington Post*, the *Wall Street Journal*, or the *New York Times*. They don't need someone in San Francisco to report on Washington politics, and the journalist in San Diego or Minneapolis adds nothing to the coverage of G8 summit meetings. No new insight will come out of the next 1,400 newspapers that didn't appear in the top ten who are best positioned to provide this niche.

USA Today may have no tomorrow. It lacks prestige and a niche.

Overall, the strategy for newspapers has become very simple: Don't bother covering areas where you're mediocre. Focus only where you can be the best.

The New Magazine: A Golden Age of Niche Publications?

Say a man gets a shrewd idea for a monthly. He's a clever writer and he knows there is an audience. Now what?

Traditionally, he needs offices, a press, a production factory, and working capital for inventory. He needs capital for shrinkage, since things get stolen and some copies don't get sold. He needs money for all the manual labor. He needs funds for the trucks and to buy shelf space, and he needs even more to launch an ad sales organization.

On top of all that, he needs to be good at all these business processes. Even if he has the world's best editorial instincts and skills, his business will die without efficient manufacturing and strong ad sales. That's why, when magazine advertising dropped by more than 25 percent in 2009,[64] a crop of titles vanished.

Indeed, magazines go bankrupt all the time.

In the new world, there's no need for capital or start-up energy. Mobile technology eliminates virtually every step between the creation of information and its arrival at the reader, so all that's required is editorial acumen and the ability to write. The modern publishing entrepreneur can focus on creating a good magazine and then upload it to the mobile store. Customers get a better product because they can download it in seconds, access it from anywhere, have it with them all the time, and junk it with a tap.

Magazines differ from newspapers. They are far more specialized, with less overlap in content. The overlap ratio among newspapers is very high, perhaps 50/1 or 100/1, while among magazines the ratio is much smaller and healthier. So we may be entering a golden age of the niche magazine. It has become much easier to distribute specialized content—the very appeal of most magazines. And, as all of the negative overhead costs fall away, a well-written, distinctive periodical may gain a following more easily.

Magazines also offer new technological possibilities that aren't feasible in the rapid-turnover world of news. For example, readers of the *The Economist* can listen to articles spoken aloud by a male voice. The magazine reads itself. And the speaker is a real person, not a computer synthesis, so this magazine has become a cross between radio and a magazine—a real breakthrough for some people, such as commuters. Instead of investing in blue-collar laborers at the printing press, the magazine can invest in a voice actor whose audio performances are embedded in the digital magazine. This richer offering might even replace certain forms of radio.

Suddenly a magazine that *can't* read itself begins to seem a tad antiquated.

A compelling and admired magazine has larger prospects beyond advertising. Suppose you are the editor of *Motorcycle* magazine. You are a hub of consumer attention, and you can be a demand channeler, sending traffic straight to music, book, and merchandise sites. Since you have the direct relationship with the customer, you can go beyond being just a magazine and become a motorcycle shop. Customers will buy from you because they know and trust you, and because you've put your reputation on the line.

You have more to lose than Joe's Motorcycle Shop down the street and, of course, much more to gain.

The Paperless Office

Paper is to an office as krill is to a whale. By 2010, the average office worker in the United States used 10,000 sheets of copy paper each year, and the national total had reached four million tons annually,[65] the weight of over a million hippos.[66]

A key reason is that people used their computers to print most of the documents they needed to read, including memos and business reports. It's safe to estimate that we've delivered 75 percent of all business reporting on paper, even if it was displayed on a computer screen first. It's like downloading an e-book, then printing it out to read.

Mobile technology will transform the "paperless office" from hype into reality. The mobile device becomes digital paper and it puts documents in your hands in a way even laptops can't. And it's much cheaper than a laptop.

A paperless office also has a better memory. Take for example, the customs form that must be filled out every time someone enters the country. What happens with all those forms? Does the government pay people to put billions of them in a filing system somewhere? It seems unlikely—once they've been viewed by customs officials, they almost certainly throw the flimsy slips away. Saving them all and filing them in any searchable system would cost far more than it's worth.

So, there's lots of cost, with little information conveyance, and no memory. It doesn't make any sense. Instead, we could have a paperless mobile system in which you could just wave your phone, have it register your identity, and tap a few times to say, "No, I'm not carrying cash and I didn't visit a farm. Now let me in." And it *would* be possible to keep such records. Filing and alphabetizing would be easy. Think of how many standard forms could be like this, at all levels of government and society.

Saving the Air

On September 10, 1994, an Australian naturalist named David Noble climbed down into one of the 500 canyons in Wollemi National Park and found a relic of a lost world. It was a stand of pine trees that scientists thought had vanished 35 million years ago. Their survival seemed incredible. Back then, the planet felt like a nature conservatory. Carbon dioxide levels were about 1,000 parts per million, whereas today they are 386, and the seas were over 450 feet deeper.[67] Antarctica enjoyed a subtropical climate, and the average temperature on earth was 77 degrees, compared to a pre-industrial average of 59 degrees.

Then the earth chilled and its cycle of ice ages began; yet somehow the Wollemi pines survived.

Now, they may start spreading again, for at least one study concludes that by 2100 the carbon dioxide levels will be back where they were 35 million years ago.[68] No one knows for sure, but it's a fact that the earth has been heating up. For instance, ships now ply the Arctic Ocean where, in 1845, ice trapped the John Franklin expedition and forced its members into cannibalism and starvation.

The death of paper will slow the warming. Paper itself is clean, but paper-*making* is dirty. Among the industries, it's the fourth largest emitter of greenhouse gases in the United States. For every book, the processes from manufacturing to retailing release 8.85 pounds of carbon dioxide.[69] There are toxic emissions, too. Each year, paper mills release millions of pounds of chlorine dioxide, methanol, formaldehyde, and hydrochloric acid into the atmosphere.[70] According to the U.S. Environmental Protection Agency (EPA), pulp and paper mills are among the worst polluters of any industry in the nation.[71]

As paper dies, more trees will live. In 2010, a one-year subscription to the *New York Times*, stacked up, weighed 286 pounds and stood over 17 feet tall—containing the equivalent of perhaps 1.7 trees. The industry itself estimates that a third of paper comes from trees felled for that purpose—the rest comes from recycled paper and residue like chips and sawdust. The Green Press Initiative states that the U.S. book industry devours 30 million trees a year, and U.S. newspapers 95 million trees.[72]

People read to obtain value from the information included on the page. They don't get value from the trees that were destroyed or the gasoline that was used to deliver the page to them. Environmental damage is a hidden cost we all pay, and mobile computers will reduce it. Soon, the burden of paper processing, distribution, and retailing will be a thing of the past. And that's just the start of the savings.

CHAPTER 4

ENTERTAINMENT

The New Universal Screen

The Disappearance of the Camera – Games: The Open Playing
Field – Movies: The Downfall of the DVD – Mobile Television –
The New Shared Media – TV and Advertising: Braving a New World

As Libyans rose up against Muammar Qaddafi in 2011, news reports were sparse at first. Readers of print newspapers heard about the killings and saw a few photos. However, viewers of online newspapers clicked the video and saw armed soldiers pouring out of pickups as stooped civilians ran for shelter and bullets kicked up dust. They saw cars ablaze, shards of glass glittering on the ground, and citizens dead in the street. Eventually they witnessed the capture of a disheveled Qaddafi himself.

Printed media displays text and photos, while video gives you the force of life. You don't just read about gunfire; you *feel* it. Mobile technology brings these experiences directly to you, anywhere, anytime.

Yet video images have long been tied to hefty objects. To catch news, sporting events, or comedy shows, we still flop down in front of the TV set at a scheduled time. To play video games, we fire up the Wii, PlayStation, or Xbox. To watch movies, we commonly drive to a theater or drop a DVD disk in the player. For pictures, we turn to cameras and the photo album.

Physical media have tied our video images down. *Casablanca* comes on one DVD and *Chinatown* on another, and many people have accumulated large libraries of these discs. But the mobile device and networks provide access to every film, and even more besides. It's *Casablanca, Chinatown, Mad Men*, Arab revolts, and home movies. It's *Angry Birds, World of Warcraft*, and chess with a guy in Chile. It's universal, and that means physical media have become unwanted baggage.

The Disappearance of the Camera

In 1841, a daguerreotype portrait cost about a week's wage, and couldn't be copied. In 2000, with film, the cost of a photo was about one dollar per shot, and perhaps 20 cents for each copy. By 2010 the initial cost was near zero, and a thousand copies could be made for free in a few seconds. The photo could be judged as soon as it was taken, so bad shots had the lifespan of a meteor. You didn't waste time, money, or the moment.

When George Eastman invented the first box camera—the Kodak—in 1888, he charged $25 for the device and $10 for developing the film. He knew where the profits lay, and the Kodak slogan became, "You press the button, we do the rest." Firms like Polaroid, Ilford, and Agfa that did "the rest" have now *gone* to rest, and Kodak and other iconic brands like Pentax and Fujifilm have buckled.

Meanwhile, novices have thrived. Casio had been a maker of electronic calculators and digital watches, and it created the basic digital camera design, complete with the LCD screen. Casio understood the new technology, and unlike the established high-end companies like Hasselblad, it had no reputation to risk. So it could learn from early kludges.

As digitization multiplied the number of photos that were taken, the Internet simplified their distribution. Where editors once had to hire their own photographers or mail transparencies back and forth to get the right images, now they could go to sites like iPhoto, sift through large libraries of stock photos, and license and upload the ones they liked.

At the same time, Flickr became a global photo album, boasting 2.7 million uploads every day.[1] To everyone's surprise, amateurs found themselves racing nose-to-nose with professionals. Costs plummeted for such images, since amateurs are much more willing to share their work, and usually are flattered by even a nominal payment.

While digitization fundamentally altered the storage and distribution of photos, mobile technology sparked the most sweeping change. The camera has all-but disappeared, having been absorbed into the mobile device itself.[2] The average person has to ask, "Why buy and carry a camera if my app-phone already has one?" Moreover, the app-phone is always with you, so you don't miss great shots, and it can email and upload them on the spot.

In May 2010, 76 percent of mobile owners used the phone to take pictures. It was the top use, more common than even texting (72 percent)[3]—an impressive stat. At this rate, as app-phones penetrate the developing world, billions of new photographers will emerge. And, when shots are virtually free, everything not forbidden will be photographed. Facebook is poised to become the new Flickr, with 2.5 billion photo uploads per month in 2010 compared to Flickr's 830 million.[4] As this occurs, your most distant acquaintances will know the intimate details your life. And so will countless strangers.

This will have sweeping implications relating to crime, privacy, and revolutions.

Games: The Open Playing Field

Evil pigs have stolen bird eggs and taken refuge in forts. So you load birds in a slingshot and lob them as bombs, battering down walls till the swine are killed.

Angry Birds was Apple's best-selling app for 2010.[5] It had interesting graphics and sound effects, and it merged strategic thinking with explosions and havoc. The small Finnish firm Rovio created it for customers to play while waiting in checkout lines, but it proved addictive and people wound up tossing bird-grenades at home, in restaurants, and at work. By 2011, people had downloaded the game 500 million times.[6] Worldwide, they were playing it for 5 million hours every day, and they had flung over 400 billion birds.[7] The iPhone version, which sold for $0.99, had brought in more than $8 million by the end of 2010.[8] On Android, where it was free, Rovio said it was yielding $1 million in ad revenue per month.[9]

Yet some observers deemed its popularity a fluke.

Zynga's *Farmville* seemed no more promising. Silicon Valley investors repeatedly failed to give Zynga founder Mark Pincus the right deal for his fledgling firm. He lacked experience in gaming, they said, and a look at *Farmville* might have encouraged this conclusion, since the gamer simply raises crops and livestock with help from friends. Yet by 2011, *Farmville* reached some 84 million active monthly users,[10] and *Farmville* farmers outnumbered real ones 110 to 1. Like Rovio, Zynga knew what it was doing, and both games showed how the field is changing.

The first video game differed from *Farmville* in just about every way. In 1960 Digital Equipment Corporation (DEC) developed the PDP-1, a pre-minicomputer that boasted a cathode ray tube. At MIT, Steve Russell had been reading the science fiction of E.E. "Doc" Smith, where black hats chased white hats across the span of the universe, and in 1962 he and his friends wrote *Spacewar*.[11] Two players used controls to move tiny spaceships across the screen and blast missiles at each other. Since then, we've seen an avalanche: *Pong, Pac-Man, Space Invaders, Dungeons & Dragons, Tetris, Doom, Grand Theft Auto, The Sims, Halo*, and thousands more.

Games are a vast domain. The research firm Flurry examined over 300 million user sessions across all apps, and found that 37 percent of them involved games.[12]

Hardcore gamers ruled this industry. They tended to be males between eighteen and thirty-four, and a market in home video consoles swiftly arose to serve them.[13] The first console, the Odyssey, appeared from Magnavox in 1972. It was analogue, silent, and graphically prim-itive, yet it was a step into a rich wilderness. By 2011, Nintendo's Wii, Sony's Playstation 3, and Microsoft's Xbox 360 had an installed base of about 203 million units worldwide.[14] They dominated the field.

Yet their reign was ending. From 2009 to 2010, console game sales fell 5 percent, largely because of mobile games, and the decline has just begun.[15]

In 2003, Nokia noticed that gamers were carrying both a Game Boy and a mobile phone and detected a market opportunity. It ham-mered together a gadget called N-Gage, a mobile phone that included a game console, a PDA, and an MP3 player. The device cost twice as much as a Game Boy Advance SP, and the speaker/receiver lay in its edge, so cynics said you looked as if you were talking into a taco. The N-Gage failed to engage the public, but it was alive with possibilities.

You could talk into this taco in a supermarket line.

Apple's iPhone arrived in 2007 with a large, bewitching screen and a potent processor, so you could play decent games anywhere, in spare moments, and you didn't need to spend hundreds of dollars on a Wii or Xbox console. There were no subscriptions, disks, or cables. The games were a snap to buy and download, and they cost less than al-most any toy.

Tablets like the iPad gave mobile gaming a larger screen, better processor, and even more arresting graphics. By early 2010, in the measure of a one-month period, 61 percent of app-phone owners had played games—the highest figure for any category of app.[16]

These mobile devices changed the games themselves. One secret of the *Angry Birds* success was micro-play. It's not just one game, it's a sprawling basket of mini-games, each playable in a few seconds. If you're trapped in a train during rush hour or dawdling in line, you can whip out the mobile phone and kill a few pigs.

The player profile has changed as well. Parents got hooked on mobile games, and the average player of games like *Farmville* is a forty-three-year-old woman. Why? As Pincus said, Facebook is like a cocktail party and social games give people something to do. In *Farm-ville*, for instance, they raise cute animals as they swap favors and bolster networks. More than 80 percent of social gamers say the pastime strengthens their ties with friends—often with friends they don't really know.[17]

The playing field has also changed for game makers. Massive public acceptance is now possible almost overnight.

Zynga introduced *Cityville* in November 2010, and it became the fastest growing game anyone had ever seen, gaining 84.2 million active users in just 30 days.[18] By comparison, the best-selling *Monop-oly*, invented in 1904 to teach Henry George's single-tax plan, then mass-marketed by Parker Brothers in 1935, has sold an average of 3.3 million copies per year since.[19] Of course, you had to play *Mo-nopoly* to get to know it, and then drive to the store to buy your own. *Cityville* went viral through electronic social networks. Its lack of price tag, easy tryout, ongoing nature, and social tentacles spurred its spread.

Alongside these changes, the profit point has shifted. In the past, the player-maker relationship ended with purchase. You bought *Mo-nopoly* and kept it forever. But today, the game itself is a loss-leader, and the money lies inside the app: the Barbie doll business model. For each user who is playing for free, social gaming companies earn $14 to $20 per free-to-play user every year by selling virtual goods, and such "in-app purchases" are now the main revenue generator among all games.[20] Thus, while John Deere sells some 5,000 tractors every year, *Farmville* sells 500,000 every day.[21]

And cost of entry has plummeted. The mobile, cloud, and social networks—the infrastructure—are already in place, and an app can ride them everywhere. All that's needed is the software. Hence, while a top-level console game might cost $18–$30 million to develop for the console platforms, with perhaps hundreds of people working for years on each game, mobile games only cost $30–$300 thousand to develop. (For example, *Angry Birds* cost $100,000.)

So anyone can take a shot at success in the new world of games.

When companies with a dash of capital can grow big so fast, incumbents may find it impossible to outmaneuver them, and change accelerates. Microsoft, founded in 1976, took fourteen years to go from zero to a billion dollars in revenue. eBay (1994) took seven years and Facebook (2004) six. Zynga (2006) approached a billion after just four years and passed long-established console game-maker Electronic Arts in market value. This quickening removes the luxury of creating a product and waiting years for it to earn a profit before moving on. A company that's too slow will be left in the dust.

Other areas of the gaming experience will be similarly impacted.

From Map to Immersion: Board games are commonly schematic maps, like checkers. They demand strategy and some, like chess, can take hours. Archeologists found the oldest known board game, *Senet*, in an Egyptian tomb dating from 3500 B.C., and King Tut played it. Its rules are unclear, but apparently the goal was to overcome evil forces to reach the Kingdom of Osiris.

Parcheesi, Backgammon, Go, and many other board games have the burnish of age and today we know *leela*, developed in sixteenth-century India, as *Chutes and Ladders*. Players sat around boards in ancient Athens, in the Forbidden City, in Versailles, and in almost every middle-class home in the West. And they all shared a common trait.

They all had more time to spare than people today.

Despite our affection for them, the market for board games tumbled 9 percent in 2010[22] and the road ahead is straight downhill. Hasbro responded with versions of *Scrabble* and *Cranium* playable in five-minute bursts, but they still faced the constraints of physical products.

With apps, every board game can move to the screen, be carried in your pocket, and you will always have them all with you. Moreover, online game centers—like Apple's—knit people together, enabling anyone to play on the spur of the moment. The digital "board" is richer graphically, socially, aurally, chromatically, dynamically, and dimensionally than the physical one. Instead of moving across a bare grid, you can step into a 3D video universe like *World of Warcraft*. The game board surrounds you.

You may still want a chess set as an *objet d'art*, but you'll likely have to purchase it at the flea market.

Odds at a Tap: Mobile technology turns the world into a casino. You can play blackjack with anyone on earth, from wherever you are, and for as long or short a time as you like. New and interesting forms of gambling can arise in such an environment. The only choke point remains the law, and the United States remains unusual in forbidding online gambling.

Ancient Roman law also forbade gambling, but it exempted gladiatorial combat and chariot races. Citizens were permitted to wager on the outcome, as they do in Las Vegas sports books today.

Mobile technology enables a more granular approach: play-by-play bets. You can wager $22 that a football team will make a first down on a third-and-ten situation. The bet and payoff occur in seconds, so the action can be fast like blackjack. But unlike blackjack, you can pick your moments. So a sports book in Las Vegas could lay odds on every play, and vastly increase the number of bets per game.

And the action isn't limited to Vegas. An individual can wager with anyone, anywhere. The 2011 Super Bowl was the most watched TV event in history with 162.9 million viewers,[23] and afterward people were sending 4,064 tweets per second,[24] another record. So, there were plenty of people who might participate in real-time wagering. And there's a social media aspect that comes into play. Everyone has an opinion about sports, and as you watch the game, you can learn what the other spectators think. Perhaps 22,000 in the stadium think he'll make first down, and 19,000 think he won't. Suddenly, the odds will be at your fingertips.

These seismic changes bode ill for console game makers. One 2010 study showed that 38 percent of iPad owners said they wouldn't buy a portable game device, and 27 percent said they wouldn't need a full-scale game console.[25] Consoles are a shrinking market, and as tablets grow more powerful, consoles will go the way of *Pong*.

Movies: The Downfall of the DVD

In the early days, a movie film was a rare and, at times, precious experience. It had one theatrical release, then vanished into the studio vaults, where some 80 percent of silent films have been lost due to disintegrating celluloid. Apart from occasional appearances at revival houses, films disappeared forever after their one big release.

In the thirty-five years since movies began to be sold on video tape, films have become a hot commodity. We can rent or own them on DVDs, or tape them on DVRs, much as we own prints of Van Gogh's painting *Starry Night*. Instead of journeying to a remote theater for a single viewing of a specific film at a specific time, we can now pluck *Casablanca* or any other movie from our DVD collection and watch it again and again, if we want. It's the difference between the *Acta Diurna* in the town square, and the print newspaper brought to your doorstep.

And now the DVD is following the newspaper into oblivion.

Local video rental stores began popping up in the late '70s, but the burgeoning chains of Blockbuster and Hollywood Video crushed most of them. Founded by a database expert, Blockbuster adapted its offerings to neighborhood tastes. It also used a novel strategy—always keep new movies in stock. This meant buying multiple copies of each title. At first, the approach seemed daring. Conventional wisdom suggested that customers would still rent a film, even if they couldn't find the specific one they were seeking. Small store owners resisted the Blockbuster approach and were reluctant to change. Blockbuster exploited this weakness.

Thus, by May 2010, Blockbuster had become the only nationwide rental chain left in the country. Yet, even Blockbuster was in its death throes, and that September the chain followed its predecessors into bankruptcy court, sabotaged by a dramatic new business model.

Netflix seemed highly unpromising, at first. The subscriber chose a movie online, Netflix mailed it to them, they watched it, and mailed it back. This incorporated a distinct lag time between the choosing of the title and the viewing. Would people tolerate the delay?

It turned out that they cared less about immediate gratification, and more about the service's low cost and vast online selection. They also appreciated the convenience. Blockbuster was perhaps ten minutes away from the average person's home, so the round trip took twenty minutes, and the customer spent perhaps twenty minutes browsing and waiting in line. Netflix offered a better alternative, wherein the customer never needed to leave the comfort of home.

Netflix delivered its billionth DVD in February 2007. And as Netflix prospered, DVDs and HD-DVDs sales went on the retreat. From 2006 to 2010, DVD sales fell 33 percent, from 958.1 million to 644.0 million.[26] By 2011, however, the Netflix's core business model of mailing DVDs was in a death swoon, and Netflix itself was doing the killing. Netflix was evolving adroitly in step with prevailing technology. It had begun to offer video-on-demand (VOD) of its entire library over the Internet. A monthly fee granted the user perpetual access, and perfect simplicity: search, click, and watch. There was no mailing and no waiting. There was no disk to carry around and, after initial setup, no time lost making a payment.

Mobile computing now offers to enhance the convenience of VOD even more. In 2010, 42 percent of mobile users who downloaded video preferred apps to a normal browser.[27] Netflix continued to evolve with the market by offering its own mobile apps for phones and tablet devices. Yet the market is still in flux and the winners will not shake out for a while. Netflix is now in a ring with much bigger competitors like Wal-Mart which offers films for $1 apiece, and the Apple iTunes Store which is becoming a high-volume distributor of video, television, and movies, with control of both the rental and sales models.

As the DVD joins the videotape in the technology afterlife, the immediate future is unclear. Studios could decide to choke off the VOD model and insist on rental and sales only. But the *ultimate* future is obvious. The simpler, more streamlined models will triumph, just as they did in rental.

Friction-free content distribution will always prevail.

Mobile Television

Rio de Janeiro cabdriver Marcelo Mendonça Guimarães likes to watch TV on his cell phone while waiting for fares.

"I actually have a digital TV in my cab, but I prefer to use the phone," he told the *New York Times* in 2008. "The reception is much better."[28] Guimarães was watching "free-to-air television," which provides regular broadcast programming that is picked up via a TV chip and telescoping antenna added to his phone. By 2010 in South Korea, 27 million people—or 56 percent of the population—viewed TV regularly on mobile phones, and some 40 million were doing so worldwide.[29] Free mobile TV service has been slow to reach the United States and Europe, where the model began arriving in 2011.

By then, it was already too late. The market has moved past scheduled broadcast programming.

Qualcomm's mobile pay-TV service, called FLO TV, reveals why. It began in 2004, and Verizon and AT&T adopted it early, but it didn't pull in the customers. To watch a show, you had to do so at the time it was broadcast. After three years it had fewer than a million subscribers,[30] and in 2010 Qualcomm shuttered the business for good.

FLO TV had an assortment of issues, but one major problem was that consumers just didn't care for *scheduled* programming—except for sports and special live events, like the Michael Jackson memorial.

"Nobody turned on their phone at 4:30 to watch show X for half an hour," Qualcomm chief Paul Jacobs said. "That was a total non-starter."[31] When TV broadcasting began in the 1940s, scheduling programming was unavoidable, but now it's a relic. Today half of all U.S. homes have DVRs—digital video recorders—and watch TV on their schedule, not the broadcasters' schedules.

We're a DVR nation.

New services from the likes of Netflix, Hulu, TV.com, and many others take the DVR concept one step further. They give you the shows you want to watch, without the commercials, more efficiently, and for a better price. They cost a few dollars a month—much less than the cable fees of $60–$70 and up—and they don't compel you to buy channels you'd never watch. These services could do to cable what the iPod did to albums: un-bundle the offering.

Moreover, like the one-run movies of the 1930s, viewers had to catch their favorite shows when they aired, or they would miss them. Today consumers can tap the memory bank. If you never saw a single episode of *Boston Legal*, you could still watch all of them on your mobile.

At your convenience.

Anywhere.

The New Shared Media

The TV audience has been graying and dwindling every year, and by 2008 the average viewer already was fifty, so an industry with a target demographic of eighteen-to-forty-nine has seen the tide go out.[32]

Viewers have shifted to the Internet. From 1948 to 2008, the three main TV networks broadcast 1.5 million hours of programming. YouTube exceeded that number in just six months, from December

2007 through May 2008.[33] In a world where an amateur videographer can make a film of a street vendor setting himself ablaze, upload it, and start a revolution, people no longer have to watch bad TV as the default.

Yet in some ways the Internet may start bringing the younger audience back.

Almost half of Americans between fourteen and seventy-five browse the Internet while watching TV, and a quarter of them instant message (IM) or text.[34] And, if they're texting as they watch TV, plenty of them are texting *about* TV. So TV can be a shared experience, no matter where the people are. For instance, viewers posted half a million tweets about the show *Lost* within twelve hours of its finale. (Many of them irate.)

Twitter and social networks offer further opportunities.

On February 12, 2011, HBO2 broadcast *Private Parts*, an autobiography of disk jockey Howard Stern. It had aired many times before, but this time Stern himself provided live tweets about the movie. He gave viewers insights into the making of the film and the events it showed, and the fact that the comments were live gave the fourteen-year-old film a new immediacy. Similarly, while you might watch a director's cut on a DVD, you can get the director's comments *live* on mobile—as well as an actor's, writer's, even a gaffer's.

Movie critics can provide real-time analysis of a movie as it runs. Comedians can mock it.

It's almost as if these people were in the room with you.

New opportunities are appearing for those who enjoy a running commentary. Tweet seats have appeared in opera houses and live theaters, where the director can comment on the production and audience members ask questions and give opinions.[35] As with vaudeville, people in the seats can talk back—albeit more discreetly.

Richer interactions are involving audiences in the performances themselves. For instance, the character Dwight Schrute on *The Office* started his own blog, and people responded as if he were a real person.

Lost had such a maze of clues that the mystery required a mass conversation on the Internet to solve. Its creators seem to have intended from the start that there be just such a group mind effort.

As shows go with us everywhere, we can experience them in new depth.

TV and Advertising: Braving a New World

Advertising is biologically ancient. The aroma of every flower is advertising that lures bees. So are the plumage of the peacock and the long tail of the widow bird. Sales ads dot Pompeii, and the taverns of ancient Uruk—the world's first great city—likely flaunted them, too.

But advertising as an industry did not begin until William Tayler opened an agency in London in 1786. He worked for publishers, as did his successors for more than a century. In 1897, the deaf copywriter Earnest Elmo Caulkins implored his bosses to use more visuals in newspaper ads. They ignored him, and he left to form the first agency that worked for companies with goods to sell.

Video is the pinnacle of visual, unrivaled in its immediacy, and major corporations such as General Mills, General Motors, and Johnson & Johnson once spent two-thirds of their annual ad budgets on television time. But as viewers dropped off, advertisers started seeking the exit. Coca-Cola funneled 94 percent of its ad budget to TV in 2000, but only 70 percent in 2006.[36]

The eyeballs turned to the Internet, and major advertisers gazed at it longingly, as if transfixed by its uncanny rules and wild possibilities. Internet ads offered better targeting and click-through purchasing. At first, advertisers focused on banner ads, pop-ups, and text boxes with flashing graphics. It was the same broadcast approach that has existed since the first bloom enticed an insect.

On mobile devices, ads started with text. SMS text messaging remains a primary way to advertise on mobile phones, and many prefer it because the customer opts in by texting a key word to a particular number, in order to receive discounts and messages. A 2009 Forrester Research survey indicated that 74 percent of marketers preferred to use SMS messaging, and 44 percent of consumers said they would rather receive requested product promotions through SMS than any other channel.[37]

Advertisers are turning to mobile apps as the next frontier. With Nike+ running shoes, the ad agency R/GA Media Group went further and created an app for joggers. The jogger put a sensor in the special slot in her shoes, and with each run it relayed to the mobile device in her pocket her speed, miles covered, and calories burned. The device uploaded this information to the Nike+ website, which held her entire running history. Moreover, she could compare her progress to that of her friends, give and receive encouragement, and join jogging clubs. She and her friends might turn it into a game. And, since others could easily authenticate that she was truly jogging two hours a week, her insurance company could reward her for it, as could her employer. Of course, to earn all these benefits, she needed the Nike shoes.

Other marketers are cranking out mobile apps at a furious pace. Proctor & Gamble's "Charmin" brand, for instance, has an app that helps the needy find clean restrooms. Some companies use games to amuse and hold the consumer, the so-called "gamification" of their products. Meanwhile, advertisers such as Starbucks and L'Oreal focus on location-based campaigns. Mobile technology is advertising's future, and vice versa.

As of 2011, TV remains caught in a squeeze. People are consuming video content more than ever before, but less and less from broadcast programming and more frequently on their mobile devices. At the same time, ad agencies are setting their sights and creative energy on the new mobile app space. TV executives have never faced a bigger challenge.

CHAPTER 5

WALLET

A Smarter Wallet and Intelligent Money

Near Field Communications: The Key – The New ID: Total Protection – The Mobile Key and Hyper-Security – The Obsolescence of Hard Cash –The Credit Card – The New Banks: A Really Big Apple – Discount Cards: Beyond Groupon – The Elimination of Robbery

As the Greek hero Perseus approached the Atlas Mountains, he seemed to enter a statue garden. Everywhere, he saw lifelike people in stone. In fact, they *had* been people until they gazed at the Medusa, the snake-haired lady with a petrifying face. Perseus had come to slay her, and he found her sleeping. Viewing her reflection in his shield so as not to be petrified himself, he decapitated her. He realized the severed head could make him the world's most powerful man, so he put it in his bag or *kibisis*, which scholars translate as "wallet."

A wallet holds valuables. Around the first century, it was the traveler's knapsack for holding food, tools, and goods for trade. In fact, a wallet and its contents were often a person's very means of survival. For coins, people used a small pouch (*bowgette* in Middle English, hence "budget") hung from the belt.

In the seventeenth century, valuables grew thin, thanks to the Dutchman, Johan Palmstruch. Holland had thrown him into debtors' prison, so he emigrated and wound up in Sweden, where he founded the Stockholm Banco. Sweden's economy was reeling after the bloody Thirty Years War, and since it had little gold or silver, it minted coins out of copper. Weighing about four pounds each, these early efforts didn't exactly qualify as pocket change, and in 1644 the Banco issued 43-pounders—ingots, essentially, heavier than the average kindergartner.

Palmstruch lobbied for paper money and, in 1661, the crown yielded. The experiment worked well at first, but in 1668 he printed too much and it all crashed. The crown ordered him executed, later commuting the sentence to life imprisonment.

Despite Palmstruch's unfortunate start, paper money began to spread through Europe.

Eventually, paper money led to the thin modern wallet, which appeared in the seventeenth century. Bills slid into it easily and it folded in half to fit into the pocket. It had a sewed-in slot to hold "trade cards," which first appeared in late seventeenth century London and became a means to give and get loans—primordial credit cards.[1] And as today's wallet carries much more—cash, credit and debit cards, loyalty cards, IDs—with mobile technologies, our thin wallet will soon become invisible.

Near Field Communications: The Key

In 1948, Bernard Silver and Norman Woodland heard the head of a local grocery chain complain about the limitations of checkout. It was slow, he said. Clerks would err at the cash register, and the machines collected no product information. In response, the pair devised the barcode, which shoppers didn't see until 1968, when a Kroger store in Cincinnati deployed it. Those familiar black-and-white stripes are ubiquitous today, and checkout has become simpler, faster, and more reliable.

One of the drawbacks of the barcode is that it requires a clear line of site in order to be read, a limitation that is eliminated with RFID (radio frequency identification) technology. With RFID, an electronic tag attached to an object uses radio waves and transfers data to a nearby tag reader. RFID tags are typically comprised of a tiny RF transmitter and receiver and can be read from several meters away without requiring a clear line of site. This technology is often used in warehouse environments for bulk reading applications, like tracking objects stacked on a pallet.

The RFID tag concept is the foundation for emerging near field communications (NFC) technology, which lets wireless devices just a few inches apart swap information. Unlike the familiar Bluetooth, NFC doesn't require a cumbersome pairing process. With its ease of use, NFC is an ideal fit for a dizzying array of potential "tap and go" applications, from swapping photos with your friends to making mobile payments or exchanging ticketing or coupon information. NFC is also a very secure communications technology, since the physical closeness of the devices and the short range of the RF signal make it very difficult for eavesdroppers to intercept data.

As of 2012, NFC is being driven to market by heavy-hitters Google, Microsoft, Visa, and American Express, along with leading mobile device manufacturers and wireless operators. NFC chips have already found their way into Android phones and the next generation iPhones will be equipped with them, too. The world will change over to NFC. Before 2020, perhaps 500 million to a billion people will have NFC in their mobile devices.[2] That is, 80 percent to 90 percent of all the world's purchasing power could come through people with handheld NFC devices.

The New ID: Total Protection

Who are you? How trustworthy are you? What are your privileges?

You know, but strangers don't. A good ID proves your identity. It could also attest to your character, and detail your licenses and qualifications.

Even in the nineteenth century, people branded, tattooed, and maimed criminals to identify them to the rest of the world. Birth certificates go back at least as far as the Roman Empire, where they could help prove age and distinguish citizens from noncitizens. Records were spotty back then, and have remained so even to today where they are still not 100 percent reliable. During World War II, U.S. law required aircraft companies to hire only citizens, yet a third of the working-age population did not possess birth certificates.[3]

As the world grew more interlinked, the leather wallet came to hold our identification, and the driver's license became a stand-in for the birth certificate. Presumably, only the owner carries the license, but to minimize theft, it also shows a photo and lists physical details like height and weight.

Online, your ID is typically validated by a password, but those are notoriously ineffective. Many people actually use "password" as their password, something even the most inept thief could crack. Even worse, with an ever-increasing number of password-protected accounts, remembering all of them is becoming virtually impossible, so many persons use the same password for every account. If a hacker learns one, he hits the jackpot.

Both the driver's license and the password will vanish as forms of ID in favor of a mobile technology equivalent. They'll be replaced with an array of biometric techniques, old and new, all of which are faster than a password and none of which require memorization. As NFC opens up the opportunity to conduct business on mobile phones, these new mobile identity techniques will secure our transactions and safeguard our data with unmatched reliability.

There are several tools that will be employed.

Fingerprint Scanning: Fingerprints have served as ID since eighth-century Japan and the Tang Dynasty of China. The pads of our fingers hold a lot of information, and each ridge pattern is unique. The genetic shuffle causes most of this variety, but fetal movements in the amniotic fluid also affect the fine detail as fingerprints are forming, so even identical twins have different ones.

In 1905, London police were working to solve the bludgeoning murder of shopkeepers Thomas and Ann Farrow. Witnesses tentatively identified two young men at the scene but felt too uncertain to name them in court. Beyond that, police had only fingerprints. Only a year earlier, the UK allowed the use of prints to be presented in court, and in the murder case they proved enough to convict the killers.

Mobile devices are driven by touch, so you, or anyone else using your mobile phone, will leave prints on the screen. Apps could read, store, and possibly take action upon those prints. For instance, a security app might lock down the device when an unauthorized user attempts access, or it might assign different privileges keyed off the user's fingerprints.

These features are only a start. With every transaction—whether buying a lawn mower on Amazon or sending money to a friend through PayPal—a fingerprint scan could call up personal or business information, authorize the deal, and streamline related processes. Scanners can also tell fingers apart. Thus, touching the screen with your little finger might call up music, whereas the index finger would bring up text messaging, and a ring finger would automatically call a spouse.

Retinal and Iris Scanning: Featured in such films as *Golden Eye* (1995) and *Mission: Impossible* (1996), retinal scanning is a relatively simple technology that has gained some mystique. Originally used in government agencies, private and federal prisons, banks, and other high-security environments, it will find its way into the iPhone and the Android platforms, using the phone's built-in camera to scan the user's retina.

The retina is the back of the eye, the "screen" of cells that respond to light. Since the retina also pre-processes images, scientists consider it part of the brain. The "red-eye" in photos is actually the retina, and scanning works because all retinas have a unique pattern of blood vessels.

The iris also lends itself to identification. The iris is the colored ring around the pupil, and it's a jumble. Every iris is different, even between the left and right eyes. The New York Police Department uses iris scans when booking suspects, and the city of Leon, Mexico, deploys iris scanners in crowded public spaces, where they can identify up to fifty people at once.

Portable Security Cameras: Your face is your prime identifying factor, and notwithstanding Hollywood makeup technology, faces are relatively hard to fake. Since mobile devices have cameras, there's an opportunity to send a face photo to your carrier as part of a security package used in every transaction. The carrier can keep the snapshots, note dates and times, and thus capture the face of anyone making a suspicious transaction.

Voice Recognition: Speech recognition is becoming common in smartphones. A user can say, "Call Mom," and the phone will obey. But speech recognition isn't voice recognition. Voice recognition doesn't try to figure out what you say, it tries to determine if the voice is actually yours. The software to identify your voice exists and would be an easy addition to the array of identify techniques available to mobile devices.

Imagine you've been pulled over while driving. The officer approaches your rolled up window and asks for your ID. You touch your mobile phone to the glass and he does the same with his own. You connect with the officer's device using NFC, and your phone prompts you to press your thumbprint and submit to a quick retinal scan. Almost instantly, the officer downloads your information from a broadband connection, confirming your identity and retrieving all relevant data.

For true security, these various techniques will layer atop one another along with your PIN. For instance, you might make the mobile device accessible only after a quick iris scan and a PIN entry. Or it might accept only fingerprint scans and, using voice recognition, a spoken pass-code. Or the phone might unlock your bank account only if you typed your PIN using your little, thumb, ring, and index fingers, in that order. Mobile security can take almost limitless customizable forms, and you won't have to remember an array of passwords.

With NFC technology embedded in your mobile phone and biometric matching capabilities, your digital ID will also serve as your key to your office, home, car, safe, and other private areas.

The Mobile Key and Hyper-Security

A key is just a way to verify identification and grant entry. Yet the traditional metal lock and key are inflexible. They can't be adjusted. It's difficult to swap in new ones. And, making copies is a cumbersome process. The limitations of the conventional metal lock and key are endless, and for years they proved to be an on-going painpoint for hotels.

The solution to those problems debuted at the Peachtree Plaza Hotel in Atlanta in 1979. The answer was the key card, a slip of plastic the size of a credit card. The first one was mechanical and used a pattern of holes, but soon key cards appeared with magnetic stripes and RFID chips.

The key card was a step up from the inflexible metal key, but the card itself will be an antique soon. Indeed, the whole process will be different. Before you even make a reservation, you'll see which rooms are available and scan videos of them, checking out the amenities and the view.

"I want to stay in room 549 at the Marriott," you'll say, and you'll get a confirmation along with an encrypted room key access code. When you arrive at the Marriott, you won't bother with check-in—you'll walk through the lobby, ride the elevator up to room 549, and tap your mobile phone to enter. After your stay you'll check out by swiping your phone against an NFC reader, deleting your room key and charging your credit card or bank.

Your mobile ID will become the key to a safe deposit box, or an office, or a floor, or a part of a floor, or a building, or an access elevator to a parking lot. You'll be able to deploy hyper-security in your own home. You'll give visitors access to your house for 48 hours, and then just turn it off. You'll control who can get into certain rooms and, for instance, keep your children or their friends out of your private study. Most people today don't have hyper-security because it's cumbersome when you try to do it with metal keys that date from Linus Yale, and trouble if you lose them.

But in a world where your phone provides the keys, and where you can reprogram them in seconds, you'll lock things down more tightly.

As hotels moved from physical to electronic to mobile keys, so did cars. The metal key had many drawbacks, but a central one was lack of singularity. The Fords manufactured in the years after 1996 had fewer than 2,000 different keys, and each could unlock thousands of cars. The door key was easy enough to fool, and once a thief broke in, he could hot-wire the car and drive away.

Then the transponder key appeared. It had a dark plastic head and serrated edges on both sides. Most people assumed the head was modest decoration, but in fact it sent a unique code to the engine control unit, so the key became one-of-a-kind, instead of one in 2,000. And since the car wouldn't start without the signal, hot-wiring became impossible. Remote controls soon replaced transponders, and enabled drivers to unlock all doors at once, from afar. Auto theft has plummeted since the days of the metal key. From 2009 to 2010 it decreased by 7.2 percent, and the National Insurance Crime Bureau credits smart keys for this decrease.[4]

Mobile keys make far more sense than previous versions. The key is on your phone, so no one can use it even if they've stolen your phone if your phone is password protected. And you can program its use. For instance, you can have a car that you or your wife can drive, but not your son. The key can use a breathalyzer test to prevent drunk drivers, and it can refuse to start the car for those with suspended licenses. The mobile car key can also serve as the glove compartment—it can store everything from the car's registration to insurance information.

Mobile car keys have enabled the Zipcar. Launched in 2000, Zipcar boasted more than 550,000 members as of 2010, and it shook up the rental car market.[5] It decentralizes pick-up locations. Cars are available at curbsides all around the major cities, not just at lots, and an app shows where they are. You pay by the hour, instead of the day, so the savings can be enormous.

In early 2011 the City of Chicago replaced more than 100 governmental vehicles with Zipcar service, and officials projected that they would see a savings of $400,000 in fuel and maintenance by the end of 2012.[6] Zipcar works because your identification is confirmed and the key is digital, so the company can send it to your mobile device wherever you are. The friction of the check-in desk falls away. The service personnel disappear.

It's just the car and you, and your mobile phone.

New approaches telescope the process even further. A company called Getaround provides an insured platform for owners to rent out their own cars. Getaround is a car rental company, without having to own a fleet of cars. The owner sets the rental fee and commonly receives two-thirds of it. The remainder goes to Getaround, which provides the supplemental insurance, handles customer service problems, and manages the payment process. If the two-thirds rental fee exceeds the vehicle wear and tear, then everyone benefits.

The Obsolescence of Hard Cash

In his book *Travels*, dated 1296, Marco Polo described a nation that used paper for money. Every year, he wrote, Kublai Khan of China manufactured "such a vast quantity of this money, which costs him nothing, that it must equal in amount all the treasure of the world."[7]

China had been using paper money since 1024, but many Europeans found the idea so ridiculous that they dismissed Polo's book as fiction. Yet, if Polo had time-traveled to the world today, he would have written about far greater wonders. His homeland Italy—and every nation on earth—now uses paper money. In Washington D.C. tourists visiting the U.S. Bureau of Engraving and Printing gaze through thick glass at the two 40-foot intaglio presses that print American currency. If visitors remained there for 24 hours, they would watch 16,650,000 one-dollar bills roll off. In the fiscal year of 2010 the Bureau printed 6.4 billion notes with a face value of approximately 240 billion dollars. Each year, 95 percent of these notes are produced to replace notes already in or taken out of circulation.[8]

It's wealth emerging out of thin air. It isn't true wealth, of course, since the more dollars they print, the less they're worth, but tourists savor the sight of blank paper transforming into a fortune, right before their eyes.

Marco Polo's fellow Venetians thought money had to embody value, like gold, but today both money and its backing have become invisible. Instead of stockpiling gold ingots, we now store wealth as on-off patterns in machines, as information. Most consumers use paper currency for trivial purchases, and coins largely with machines, and both are headed for the grave.

Like the printing press, cash was once a great boon. Herodotus, the ancient Greek historian, guessed that the first coins appeared in Lydia—in what is now Turkey—perhaps around 620 B.C. Coins had three spectacular advantages over barter and commodities like cowry shells. They lubricated trade by being widely accepted, they stored value conveniently (in the bank or under the mattress), and they measured it in clear units. Together, these virtues quickened the flow of wealth and enriched Lydia. Coins led to Lydia's vigorous expansion to the point that one of its kings, Croesus, became a byword for fortune, as in "richer than Croesus."

Paper money made commerce swifter. Bills were slender and almost weightless—U.S. notes are 0.0043 inch thick and each weighs 1/30 of an ounce.[9] So people could carry them easily, which is crucial for currency, since its ultimate purpose is for it to change hands.

The Bureau of Engraving and Printing has manufactured all U.S. currency since October 1, 1877. It began with six people down in the Treasury Department basement printing cash with steam-powered presses. Today, the Bureau's presses consume 18 tons of ink every day.[10] It has 2,300 employees spread across 25 acres in Washington, D.C., and others staff a printing plant in Fort Worth, Texas. It has 160 security guards and employs ultra-high security equipment. The Bureau also employs about a dozen engravers, who—using scalpels unchanged for over two centuries—etch an intricate webs of lines, letters, and numbers into the chromium-coated plates. They work in mirror image with powerful magnifying glasses, and if one hand slips, the plate goes into the trash and they start all over.

It can cost months of labor.[11]

Despite its light weight, however, cash has become an economic burden. Transactions are time-consuming. Currency gets soiled, torn, and wears out quickly. The paper in U.S. currency comes from the scraps that didn't make it into your blue jeans,[12] but denim itself is much more durable. The average five-dollar bill lasts for sixteen months, the ten for eighteen months, and the single for twenty-one, so we actually spend money to take money out of circulation.[13] Each year, the Federal Reserve bleaches and shreds 7,000 tons of worn currency, some 10 percent of which goes to roof shingles, and the rest to the dump, where it takes up otherwise usable space.[14]

Paper money requires complex tactics to thwart counterfeiting. In the thirteen American colonies, some printers infused the paper with mica particles to thwart replication. Naked skin seems to challenge forgers, and a few nineteenth-century banknotes showed discreet nudes. Even so, around the time of Andrew Jackson, about a third of the currency in the U.S. was counterfeit.[15] In the late 1980s, North Korea began spewing out supernotes, fakes so superb the U.S. government redesigned its currency for the first time since 1928. Faces grew in size, since expressions are hard to replicate, and new watermarks, security threads, and color-shifting inks appeared.

Currency is hard to trace, so it aids crimes like drug trafficking and theft. Then we all pay—in tax dollars—for prosecuting and jailing the offenders, for probation officers and halfway houses, and ultimately for a loss of economic productivity.

Coins have their own drawbacks. While they do last much longer than paper money—about thirty years—and some 6.4 trillion coins were circulating in the United States in 2010,[16] they also involve hefty manufacturing costs. In October 2011, the penny contained metal worth 50 percent more than its face value, and the nickel, 101 percent more.[17] Coins have gone from a basic standard of exchange to an annoyance. Totaling them up takes time at the checkout counter, and the clerk has to make sure there is always enough change on hand. The merchant has to sort and total the coins and ferry them to and from the bank, where employees spend further time dealing with them.

Las Vegas was once the world capital of coins. Its casinos strained local banks by ordering millions of dollars each week—mostly nickels, quarters, and half-dollars—all hauled in by guards driving armored trucks. For decades, casinos paid not only for delivery, but for fixing slot machines when coins jammed them, for staffing coin-counting windows, for filling and emptying machines, and for managing currency inventories.[18]

By the late 1990s coins had become such a burden that the gaming industry sought escape. Ultimately, it adopted "Ticket-In Ticket-Out" (TITO). The player inserts a bill or a ticket into a slot machine, chooses an amount to bet, and pulls the lever. The machine tracks the total and, if the gambler is owed anything, prints a new ticket that can be redeemed for cash at an automated kiosk. By one estimate, TITO has saved casinos as much as 40 percent of their costs.[19]

As TITO proves, money doesn't have to be physical. Currency can become pure information, a matter of numbers swapped back and forth. If I buy from you, my account deletes $20 and your account gains $20. Coins, paper bills, even cowry shells are all just the physical tokens of those accounts. If we trust the accounts themselves, we don't need the tokens.

As digital cash eliminates the tokens, it will eliminate the giant intaglio presses and the engravers and the security guards and the ink and the paper. It will stop the trucking of dead cash to the dump. It will end the need for security threads and color-changing inks, and the worry about supernotes. The average bill costs 9.6 cents to manufacture today;[20] tomorrow it will cost virtually nothing.

Digital cash will also enable money to move instantaneously across the face of the planet. It's friction-free, and more than just a replacement for paper money; it has multidimensional negotiability that can be programmed. For instance, its use could be limited to specific items, so a parent could give a child funds that could only be spent on school supplies. It might only be spent before 9 p.m., or within two miles of home. Digital cash allows customization in any dimension you can contemplate.

Take city parking for instance. Mobile devices make parking meters themselves smarter. Instead of feeding coins into a slot, with NFC you would tap the meter with your mobile phone, and a central server would register the payment. Tap it when you arrive and tap it when you leave, and you would only pay for the time you actually used.

Meters could adjust to make traffic flow more fluidly. In market terms, the ideal parking rate eliminates the search for spots while keeping all of them filled. It achieves what economists call "price equilibrium," the perfect match of demand and supply. Yet demand varies constantly, and the ideal parking rate is always in flux. Via mobile technology, meters could measure their own usage, measure the usage of the other meters in the city, and know which spaces are the most valued at different times during the day. They could automatically charge more at peak times and in crowded areas, declining with distance from a popular boulevard. Meters would therefore make all spaces appropriately desirable.

Mobile technology makes this all possible.

And the meter becomes its own meter maid. The central database knows if you exceed the time limit and can issue citations automatically. Fines can also be progressive. The longer a driver keeps others from the space, the higher the fee. Cities save on personnel and billing, earn more revenue, and manage congestion.

Who still relies on cash? There are three significant populations.

"It is only the poor who pay cash," writer Anatole France said in the early twentieth century, "and that's not from virtue, but because they are refused credit."[21] Low-income consumers make far more small purchases, and they inhabit a cash economy. But this same demographic has embraced mobile devices as an inexpensive way to tap the Internet. This will place the equivalent of an ATM in the palms of their hands, so it won't matter if they have been refused credit.

Anonymity is another major reason for the use of cash today. Cash is hard to trace, so drug traffickers love it. One reason the United States prints no bill higher than $100 is to prevent smugglers from cramming much more than $400,000 into the typical briefcase. Likewise, terrorists prefer cash and commonly stage robberies to get it—as with Joseph Stalin's spectacular 1907 heist of $3.4 million in Tiflis. Tax evaders like cash, so in 2010 struggling Greece banned all cash deals greater than 1500 Euros.

You can't hide transactions on the mobile device. If you could, cash would disappear, and ATMs would follow.

Cash still prevails in emerging markets. China has 530 point-of-sale terminals and ATMs per million people, and most of them lie in cities. By contrast, the United States has 10,000 per million. The 750 million people of rural China pay almost entirely in cash—overall, the Chinese use cash in 83 percent of their transactions, compared with 21 percent in the United States.[22] This dependence on physical currency makes it harder for buyers to buy and harder for retailers to sell. And China is hardly alone. In India, Indonesia, Brazil, and Mexico, more than 95 percent of all consumer-initiated transactions are cash-based.[23]

The problem is hard for westerners to imagine. People in emerging markets have to journey into the cities to deposit or withdraw cash, or to pay bills. Once at the bank or post office, they have to wait in line, commonly for two hours or more. These trips can take hours and waste half a day's potential salary.

With access to a mobile device and digital cash, these sorts of transactions could occur in an instant, without the wasteful friction. On the mobile phone they could buy goods more cheaply and from anywhere on earth—their markets would go from local to global.

Despite the continuing reliance on physical currency, mobile technology is spreading quickly in the emerging world. India and China added 300 million new mobile subscribers in 2010,[24] about the total population of the United States. Digital cash, throughout the world, is inevitable.

Today, 90 percent of the world has access to a mobile network—more people than have access to clean water.[25] These individuals will leapfrog into the twenty-first century, at least on a financial basis. Digital cash will open new markets and accelerate the pace of entire societies.

The Credit Card

You probably have a credit card in your wallet. If you're like the average American, you have four, and one in ten of us have more than ten cards. A stack of all the 610 million cards in the United States would reach seventy miles into space, taller than twelve Mt. Everests piled atop one another.[26] More than 175 million people in the United States have credit cards.[27] They are an international currency, and Visa is accepted in almost every country on earth, at more than 20 million merchants.[28] We use them daily in almost every kind of store, and hardly give them a thought.

The automobile helped to bring household credit into the light. Henry Ford's Model T, the "car for the great multitude," cost $360 in 1916, about half a year's income for the average American.[29] It was more than most could afford up front, but a system was devised allowing the amount to be paid on the installment plan. So in 1924, Americans bought almost three out of four cars on "time."[30] In fact, the installment plan proved as important as the assembly line in creating the mass market for autos.

Not surprisingly, an array of household lenders soon appeared, from General Motors Finance Company to department stores, credit unions, and remedial loan societies.

Credit multiplies wealth. A depositor puts $30,000 in the bank, then the bank loans it to someone to buy a car. At that point four entities lay claim to that $30,000 value: the depositor (who earns interest on it), the bank (which earns higher interest), the purchaser (who has the car), and the car dealer. If the car dealer buys a painting with it, and the artist uses the money to lease a loft from a woman who puts the money back in the bank, then that $30,000 has multiplied six times.

And the bank has it twice.

The Bureau of Engraving and Printing seems to be creating wealth when it prints money, but credit does a far better job of it. So that which expands credit enriches the world.

One evening in 1949, Frank McNamara, the head of the Hamilton Credit Company, was dining with friends and realized that he'd left his wallet at home. To pay the bill, he had to call his wife and ask her to bring it to him. The embarrassing incident illustrated to him a basic limitation of cash: you could spend only what you had on hand at a given moment, no matter how much money you had in the bank. Hence, he formed the Diners Club—the first credit card accepted by a wide variety of merchants.

McNamara and his friends returned to the restaurant a year later to reenact that evening, and they all paid with cards. In the credit industry, this meal is known as the "First Supper." That's the story, anyway, and it may even be true.

But the upshot was undeniable. In 1951, the Diners Club handled about $35 million in transactions. Seven years later, in 1958, it handled $465 million, with a gross profit of about $40 million.[31]

Other companies took note, and cards proliferated. Early entrants included American Express, which introduced its Green Card in 1958, and Hilton, which broadened its company card into Carte Blanche in the same year. In 1966, both MasterCard and BankAmericard—which became Visa in 1976—made their debuts, and today they dominate the market. Ironically, Diners Club today has just 0.5 percent of the credit card pie.[32]

These universal cards made credit a one-stop affair. Instead of maintaining numerous store-by-store accounts, customers could carry just a few cards that they could use everywhere. Credit cards not only made paying faster and more convenient, but they enabled new service arrangements, like Internet purchases, and provided security to businesses like hotels and rental car agencies. Hertz wants to know it can bill you when you drive off the lot with a costly vehicle.

Credit cards reign supreme in our wallets. In 1970, only 16 percent of households boasted credit cards, and the average household spent $564 a year with them.[33] In 2010, the Census Bureau says, the estimated annual credit card purchase volume reached more than $2.7 trillion, or $25,714 for each household. Total credit card debt in the U.S. at the end of 2010 was $800.5 billion. That was down from $974 billion in pre-crash August 2008, but it's likely to bounce back.[34]

We live on credit. Yet the cards themselves cause market friction.

At $1.10 a piece to manufacture, in the United States alone credit cards cost $671 million to make, and they need to be replaced every three years or so.[35] They also require card readers. The companies send you your balance only once a month, so it's easy to exceed it. Companies go to great expense to make the cards secure, yet thieves can steal the cards, take the number from a used charge slip, and conceal devices in ATMs that will grab both your credit card number and PIN.

Excluding staffing and management costs, U.S. credit card fraud costs at least $16 billion dollars a year.

Digital credit eliminates all that friction. The expense of making cards and manufacturing readers drops to zero. There's no physical card with an expiration date. Since you are perpetually connected to the Internet, you have access to your balance all the time. With NFC and biometric protections, digital credit transactions are inherently secure. Unlike the wide-open magstripe on a physical credit card, apps can require special authentication codes for each transaction. The potential impact is so great that if card companies gave everyone in the United States a free mobile device, they would earn back the money through the drop in fraud alone.

As was the case with digital cash, when packaged as mobile software, credit can be much more flexible. If you want to let your son use your credit account, you can limit his spending. You can restrict it to a certain store, and temporarily replicate the old one-company card. You can exercise budget discipline so that you can limit various types of purchases to stay within predefined bounds.

This isn't just a scenario for the most wealthy nations. Digital credit will become available to people in the developing world, too, perhaps first as "micro-credit" involving business and personal loans of a few dozens of dollars.

"The American standard of living was bought on the installment plan,"[36] historian Daniel Boorstin said, and electronic credit may bring affluence to dusty corners of the world that hardly know its meaning.

The New Banks: A Really Big Apple

In 1317, English churches wanted to move a large sum of cash across Europe to the Vatican, but highwaymen patrolled the roads, so the journey was perilous. So the London branch of the Italian bank Pe-ruzzi and Bardi took the cash and sent Rome a receipt, called a "bill of exchange." This bill of exchange was like a check: valueless to thieves.

When it arrived in Rome, the papacy presented it to the bank branch there, and collected the money. The primary purpose of banks was security.

Today, when cash disappears and ATMs get recycled and the seventy-mile-high tower of credit cards collapses, what will be the role of banks?

Banks will remain, for the very reason they arose in the first place, se-curity of funds and transactions, but in a distinctly different form. Suppose you have $100,000 in digital cash stored on your phone. Lose your mobile device, and the digital cash is gone. The good news is that, thanks to your mobile phone security precautions, the phone finder can't get access to your money. Unfortunately, neither can you. To rescue your digital cash, you'll need to have a record of it on an-other server, and that will be a bank. In the new world, banks will keep your monetary value as records stored in a *digital* vault.

You'll also need a bank account for other reasons. You'll need it to handle all the debit and credit transactions you conduct every month. You'll also want it as a credit reference, to vouch that you are trustworthy.

Banks will stay, but branch banks will collapse. In the old world, if you wanted a mortgage, you went down to your neighborhood bank branch. You walked into a building where you found tellers behind bulletproof glass, offices, electricity, and heat, and you spoke to a well-paid individual with an excellent health plan and a company car. And you *paid* for all of that overhead, when it was incorporated into the mortgage rate. So, instead of paying 450 basis points, you paid 480 basis points or, for all you know, you paid 20 percent more in interest.

Thanks to those expenses, your house might end up costing $2,000 a month instead of $1,800.

With digital credit, you will use your mobile device to visit an on-line bank in the Netherlands. If the Dutch bank will give you a mort-gage at an interest rate of 4 percent, and your local bank wants 5 percent, then staying local may not be the best option. We're rapidly moving toward an era where the impediments to globalized finance will be regulations and laws, but not the laws of physics.

Technology companies such as Amazon are moving into financial services, and this likely represents a major shift in our economy. Amazon can make this leap because the old prerequisites—ATM networks, brick-and-mortar bank branches, loan officers, credit cards, and pro-cessing technology—are no longer relevant. As they disappear, the barriers to entry drop, and new companies move in. Banking becomes a relationship business, and Amazon and Apple have plenty of rela-tionships they can leverage. They'll naturally view finance as a line extension of their existing businesses.

Apple has more direct customer billing relationships than any other company. The iTunes Store boasts 200 million users[37] and every one of them has posted a credit card, so Apple can directly debit 200 million people. That's more than Amazon, more than Wal-Mart, more than any telephone company, and more than any other bank.

And Apple is in an intriguing position. If it can debit the credit card, it can *credit* the credit card. Since every iPhone is connected to iTunes, it is effectively a credit card. Thus Apple could bypass Visa. It could also become a much larger consumer bank than Bank of America, because new people join its network every day. Continuing on the current course, and before long, Apple will have 500 million consumers whose accounts they can access. They're credit card accounts today, but they could be-come direct bank accounts tomorrow. If it happened, Apple could let you touch your phone to anyone else's and transfer funds.

Every single iPhone could become an ATM to every other iPhone.

Of course, such a move by Apple or Google—toward becoming a global, unregulated bank—won't happen without a cataclysmic struggle. The traditional banks and credit card companies may be reluctant to join an NFC network or share money with a tech firm. Nevertheless, they're going to lose their value-added features, and hence lose influence and economic power. So in the end the struggle will exhaust them, and they will capitulate.

The harder they resist, the worse they'll fare, since they'll fall further behind.

Discount Cards: Beyond Groupon

What else do you have in your wallet? You probably have discount cards of some sort—either coupons or loyalty cards. One rewards you for coming in, the other for coming back.

Today, we're witnessing a strong upswing in "extreme couponing," full of self-declared "super-couponers." Partly because of the 2009 recession, coupon use hit an all-time peak in 2010—when U.S. consumers redeemed 332 billion coupons worth $3.7 billion. But the coupon world is changing swiftly. Though some 87 percent of these coupons were the old clip-out kind, *digital* coupon offers increased by 37 percent, and striking new kinds appeared.[38]

Most coupons are splashed across the Sunday newspaper, but they arose in the more sophisticated realm of finance. In the early nineteenth century, if you loaned money to an organization—that is, bought its bond—you received a certificate with little squares at the bottom. To collect your interest, you cut these squares out and presented them at the organization's office. The name for these squares came from the Old French *colpon* or *copon*, meaning "a piece cut out."

A coupon was a claim on value back then, and it still is today.

It was the Coca-Cola Company, based in Atlanta, Georgia, that made the first use of the coupon technique. In 1888, when the company was almost a newborn, executive Frank Robinson heard that a customer had refused to try the drink. In response, he printed free drink coupons, redeemable at soda fountains, and mailed them out to names pulled from the Atlanta phone book. It was a smart idea, for a coupon succeeds when the recipient continues to purchase the product. Thus the cocaine-laced beverage ensnared repeat customers.

The coupon also achieves what economists call "price discrimination." One price doesn't fit all—some people are willing to pay more for your product, and others less. You can still make a profit off the latter group, however, if you can identify them and extend a lower price *only* to them. Coupons enable you to do that.

But coupons take time and effort, and those who are willing to pay more for a product are less likely to invest the energy required to seek out the coupons, clip them, and remember to bring them to the store. What's more, for some people they are an embarrassment at the checkout line and many coupons are for products consumers may neither need nor want at that time. So most coupons go unused. In 2002 alone, U.S. companies handed out 248 billion coupons worth more than $221 billion, but consumers redeemed just 1.5 percent of them.[39]

Paper coupons harbor hidden costs and inherent friction, too. Companies have to design them and then pay for space in the newspaper, or for independent printing and mailing. When you present them at checkout, the line comes to a halt, slowing commerce. Others wait as you pull them out, then while the clerk looks them over, scans them, and puts them away. The store itself has to manage them, turning them in for credit. And some customers buy only the discounted items—nothing else.

Digital coupons distributed via mobile phones strip away these costs, one by one.

No one pays for printing or newspaper space—customers simply download their coupons and redeem them via NFC at the cash register. It's fast, it's easy, and it's invisible to others, so the customer's vanity remains intact. Retailers and services businesses maintain a record of the transactions, and place each customer on a target list. They'll craft special offers and prices for each individual, and produce dynamic, personalized coupons that reflect the customer's tastes and buying patterns.

Digitization has enabled the *group* coupon, too. More than fifty million Americans have joined one of the hundreds of sites that have taken flight. They're all basically the same, so eventually the field will collapse, and a few companies will dominate. To participate, you sign up and give a firm such as Groupon or LivingSocial your zip code. The firms send you a geo-targeted email, Tweet, or Facebook message every day offering a discount of 50 percent to 90 percent, redeemable at a local business such as a Chinese restaurant. The rub is that nothing happens unless enough people accept the deal. If they do, you and everyone else fills up the restaurant that night. It's a rugged experience for the restaurant since, after honoring a 50 percent discount, the establishment must split the remaining 50 percent of their revenue with the coupon firm.

In the end, the retailer keeps only 25 percent, hoping that some of the couponers will come back.

That lost 75 percent discount is a big gamble. If the restaurant's normal gate for the evening is $28,000, it pays $21,000 in discounts to bring in $7,000. It's also an impersonal, brute force approach that may fill the seats, but doesn't necessarily foster customer loyalty.

For that, the restaurant might do much better with a loyalty program.

The loyalty card is the most common form of loyalty program today. The basic concept is that the customer receives a card, and the retailer punches a hole in the card or scans it with each purchase. After a certain number of punches or swipes, the loyal customer receives something for free.

More than 75 percent of consumers today possess at least one such card, and in 2003 U.S. companies spent more than $1.2 billion on these programs.[40] Unfortunately, many of them simply fail. If all competitors use similar cards, the program won't work. If the reward seems difficult to achieve, too insignificant, or the rules keep changing, the customer won't bother. In one study, 43 percent of customers indicated that they had never used a loyalty card at all in a year, and 36 percent had used it less than six times.[41]

A loyalty program can also work too well, as in the case of the "pudding guy."

Healthy Choice food company offered air miles for customers who bought its pudding cups, and David Phillips, a bargain-loving engineer from Davis, California, noticed a loophole. He spent $3,140 on the desserts and earned 1.25 million air miles, estimated to equal the value of $25,000 to $75,000 in free airplane tickets. This meant that he had to do something with 12,000 pudding cups, but he just gave them to charity.[42]

For all of their potential benefits, loyalty cards waste just as much space as any other plastic card in the wallet. They tend to proliferate, since each business issues its own card, and before you know it, your wallet is bulky and less organized. Eventually, you may stop joining loyalty programs altogether, simply to escape the clutter.

The digital wallet can do for loyalty cards what it did for credit cards. Free of physical constraints, the wallet can hold a galaxy of cards that you might load into the phone using the camera to scan the barcode on each one. As these accumulate, you can keep them organized as easily as your music playlist. You will accrue your loyalty points at the store using NFC communication, or simply by displaying a barcode on your screen to be scanned at checkout.

A single universal platform for loyalty cards would be even better. Technologists at MIT are developing a universal software platform for loyalty cards called Eclectyk that obviates the need for each individual company to produce their own cards, their own applications, and their own systems for distribution and maintenance. This can be costly and is repeated over and over again. In one study by McKinsey, large retailers paid up to $30 million in the first year, and $5–$10 million each year thereafter to develop and maintain their own loyalty programs.[43] Eclectyk would provide a single platform that any loyalty program could use, thus cutting out the development and maintenance costs.

Digital loyalty cards also make it easy to reward customers who act as recruiters. They make it possible to record (and reward) a customer who brings a friend into a bar, restaurant, retailer, gym, and such, and signs them up for the loyalty program. Accurate record keeping allows businesses to be creative in construction of their programs, adjusting rewards based on the timing or frequency of the purchases. Instead of using the "Groupon" discount model, and offering sweeping discounts all in one night, a restaurant or club could reach out to specific consumers—highly influential among their social groups or micro-celebrities—for targeted promotions. A software-based mobile loyalty system could keep track of which micro-celebrity invited whom, and which of the invitees accepted the offer so the promoter can be rewarded.

We might also see a new set of pyramid marketing concepts like the Amway or Mary Kay up-sell. You sign up five people, each of them signs up five people, and so on. As the person who began the recruitment drive, you might receive 25 percent of your first line sales, 10 percent of the second, 5 percent of the third, 2 percent of the fourth, and 1 percent of the fifth. No one could effectively craft programs like this using physical media and cards, because the record keeping would be far too cumbersome.

The advantages of transforming loyalty cards into mobile software go beyond those of mobile credit. It gives the consumer less to carry, less clutter to manage, and greater convenience in using their "cards." Businesses will see less loyalty fraud and more consistent usage, as well as the ability to create many new kinds of campaigns. These benefits will drive out the coupon and loyalty world of paper, plastic, and "Groupon" and replace it with loyalty systems that run on universal platforms on the mobile device.

The Elimination of Robbery

On January 13, 2011, a pickpocket stole $5,000 from the wallet of a man working at an auto parts shop. The victim had gathered the money for a car auction later that day. The thief approached the victim to discuss the price of a certain auto part, and then made her move. "She came close to me and said, 'Please give me a good price,'" said Javier Medina, the victim. "I haven't been able to sleep since this happened."[44]

The advent of the thin wallet, introduced in the 1600s, proved a boon for thieves. They no longer had to cut a purse free and dash away with it. Instead, they just lifted the slender billfold and disappeared long before the victim even noticed. Picking pockets is rather easy. Children can do it, and in nineteenth-century New York there flourished a world of pre-pubescent thieves. The crime was literally a magic trick. Narrow the mark's attention—often by bumping into him—and he won't feel his wallet being removed. "A big move covers a small move," as conjurers say. With every stolen wallet, the child gained a median $10.30, at a time when working men averaged $5 to $10 per week.[45]

In 2009, some 133,000 cases of pickpocketing and purse snatching took place in the United States.[46]

And those incidents led, in turn, to 43 percent of the instances of U.S. identity theft. Identity thieves stole $31 billion from Americans in 2009, and businesses worldwide lose up to $220 billion a year to such fraud.[47] The harm isn't just monetary, either. About half of the victims were denied credit, about half got harassed by collection agencies, almost a third couldn't erase false information from their records, and a twelfth suddenly had criminal records that they couldn't clear—all because of stolen ID.[48]

"I was a nervous wreck," Donna Pendergast said in 2009. She was the victim of an identity theft ring called Cannon to the Wiz. "All I could think of was a house being [bought] in my name."[49]

For every crime, the cost includes the value of whatever was stolen plus a long tail of additional costs, too: labor to catch the criminal, judicial expenses, cost of imprisonment, salaries for probation officers and counselors, insurance claim processing, psychological harm, lost productivity, the price of better security, and more. Estimation methods vary, and according to an average of three recent studies, each robbery or armed hold-up costs $67,277, each household burglary $13,096, each motor vehicle theft $9,079, and each larceny (other theft) $2,139.[50]

By Department of Justice figures, American society paid over $33 billion for these crimes in 2009.[51] We've tried everything to stop them. Pillories and public whippings failed. Imprisonment has failed. Rehabilitation has failed.

What if we remove the opportunity, instead?

In fact, the decline has already begun. Greater security and public awareness have helped reduce identity theft and its damages. Cases of robbery fell about 35 percent between 2000 and 2009, and instances of other property crime, about 28 percent.[52] Why? The answer is complex, but technology is certainly a factor. Not only are spy cameras ubiquitous, but thanks to cell phones, everybody has a panic button on their hip. If a robber mugs a man on the sidewalk, a witness can call 911 in seconds. That person can also photograph—perhaps even video—the assailant, and send this evidence at once to police.

With the digital wallet, pickpockets and muggers will lose incentive altogether. Crimes like burglary and car theft arise from the availability of physical targets like keys, ID, paper, and cash. With the digital wallet, there's no key to steal. There's no physical ID to steal. There's no cash to steal. There's no credit card to steal. There's no wallet or purse to steal.

You might steal a mobile phone, buy you can't grab software.

What about digital robbery? If a thug puts a gun to your head and demands that you identify yourself to your mobile phone and transfer thousands of dollars to his account, you'll do it. But there will still be an electronic trail, leading to those digital dollars, so police can track him down. Thieves will handle digital money at their peril.

Suppose a pickpocket stole your mobile device itself. The digital wallet contains far more wealth than a leather wallet—could she transfer money to herself, steal your personal information, or plant key-loggers or other malware?

The answer is a string of no's.

First, she'd have to overcome all the biometric security. If the mobile device won't turn on without an iris scan, it's junk to a thief. In addition, built-in measures could shut it down when an unauthorized user tries to pry. If the mobile device takes her picture and sends it to police, the device becomes radioactive to the thief. Moreover, GPS makes it traceable, so the thief is a flare on the map. And, as mobile phones become increasingly inexpensive, people have less reason to even steal them for their resale value.

What use does a pickpocket have for a wallet that is a vault to open and hazmat to handle? If a crime is pointless, people don't commit it, and if it automatically incriminates them, they don't even dream about it. As a result, we lasso a whole set of societal efficiencies. Jail cells open up. Court dockets grow less crowded, and probation officers have less work. Fewer people become victims, neighborhoods grow safer and more prosperous, and the quality of life improves.

This opens up another aspect of mobile technology—one that may cause many consumers to pause and think. If using a stolen mobile wallet is no longer anonymous for a villain, then it won't be anonymous for its rightful owner, either. We have to expect all of our purchases to be documented, publicly linked to us, and that information used by companies to better understand our purchasing habits.

In George Orwell's *1984*, Big Brother was pretty clear about what he was doing. But many today fear corporations are snooping on them in secret and will abuse the knowledge they collect. In a 2011 poll by Accenture of 1,100 "tech-forward" people worldwide, 73 percent said that using a mobile phone for payments made them worry about their privacy.[53]

This anxiety is misplaced and deserves inspection.

Companies generally collect information about us in order to serve us better—that is, to learn what we want, and offer it to us. If they succeed, we rarely cry foul. For instance, who objects to Amazon's use of data to make book suggestions? And yet in doing so, Amazon has analyzed your purchases. It knows far more about your literary tastes than Barnes & Noble, or perhaps even your friends.

Nevertheless, you benefit.

Companies also have strong incentives not to abuse the information they gather. First, if such activities were discovered, it would drive customers away. Big Brother didn't need to worry about that—he wasn't trying to sell people shoes. But Google, Apple, and most companies care very much about consumer good will. Markets are intricate feedback webs, and consumer backlash can cripple a firm. What's more, abuse would attract media attention, and perhaps even government reprisals.

Hence, most commercial violations will be curtailed fairly quickly and new technology will appear to overcome people's other fears.

The scale of our concern can be measured empirically, by gauging the choices consumers make; almost always we choose greater service for more information. Credit card companies know what we buy, but we use the cards anyway. Cell phones companies can triangulate our location and even audit conversations, and yet people have bought and used five billion of them. The benefits are compelling, concrete, and immediate, and vastly outweigh the concerns.

Cases of companies abusing their access to consumers' privacy are so uncommon that it is hard to think of any. The district attorney is much more likely to violate your privacy. Firms are regulated; governments are not. They do the regulating. Governments lack the checks and balances that hem firms in. Indeed, governments can demand eavesdropping.

The real concern doesn't stem from the tracing of purchases enabled by a mobile wallet. There's a much broader issue: the erosion of privacy brought on by the Internet. The problem lies in the very nature of the Internet where everyone is a publisher, producing content and sharing it. But the record is permanent, and the audience is global. So a photo that's made public—even if it's completely out of context and thoroughly misleading—can forever become a part of a person's life record.

The law hasn't caught up with this problem.

Modern legal concepts of privacy originated in 1890 when Warren and Brandeis, two Harvard professors, wrote of a "right to be let alone." In 1960, William Prosser, the most influential tort scholar in U.S. history, argued for four different privacy claims, and more than half of the states adopted them. But Prosser lived in a pre-mobile world. Three of his torts—false light, intrusion upon seclusion, and public disclosure—apply only to deeds that occur behind closed doors and outside of the public's view.

The fourth, misappropriation, typically addresses misuse of a name that holds tangible commercial value—Lady Gaga, for example—and hence affects mainly celebrities.[54]

In 2010, an Italian court convicted three Google executives of violating the privacy of an autistic boy. They were charged with letting others post a video of him being bullied. Perhaps for this reason, Google has since limited its "Goggles" feature—which lets you snap photos and get information about what you've shot—so that it can only be used with physical items and not with people.

The legal system might be beginning to recognize the problem, but technology is quick, and the law is slow. Technology thrives on breakthroughs, while the law relies on historic continuity. So while the legal system might eventually adapt, it's not at all clear when or how.

CHAPTER 6

SOCIAL NETWORKS

A Mobile Social World

The Development of Mega-Cities – The Rise of the Social Networks –
The Evolving Social Phenomenon

FIGURE 6.1 Social networking sites like Facebook,
Twitter, and YouTube were the organizational engines
behind the Arab revolutions at the start of 2011, a time
that came to be known as the Arab Spring.

On March 3, 2010, a street vendor named Abdesslem Trimech set himself ablaze in Monastir, Tunisia. Officials had harassed him until he had reached his breaking point. He wasn't the first to protest in this drastic way, but like most self-immolation events, few people ever knew it had happened.

Nine months later, on the morning of December 17, 2010, another Tunisian—Mohammed Bouazizi—was selling vegetables from his cart in dusty Sidi Bouzid. A policewoman asked to see his license. Like most peddlers, he lacked one, so she commandeered his cart, demanded the equivalent of a day's earnings (about $7), and slapped him. Humiliated, he went to a government building to complain, but officials refused to see him. So he walked to a gas station, bought a canister of fuel, and by 11:30 a.m. returned to the building. Standing in traffic, he drenched himself.

"How do you expect me to make a living?" he cried. Then he set himself on fire.

As with Trimech, Bouazizi's deed might have gone unnoticed to the world at large, but two men used their cell phones to make a video of the illegal protest. Then they posted it on Facebook. The Al Jazeera news agency regularly scanned Facebook for newsworthy videos, and that evening they broadcasted the video.

Later that day people took to the streets "with a rock in one hand, and a cell phone in the other," as one put it. Before Tunisia's president, Ben Ali, knew it, his country was rising against him, and viewers across the world watched it on TV. Rebels stayed in touch through Facebook, which the government hadn't censored. A third of Tunisia's population had access to the Internet, and 50 percent of its unemployed workers were college graduates,[1] so channels were open, and plenty of educated people understood them. President Ali tried to crack down, but on January 14, 2011—less than one month later—he fled the country.

A string of revolts followed, the likes of which hadn't been seen since Europe in 1848. They flared across the Arab world, fueled by Facebook and Twitter communication feeds.

Mobile technology has made us hyper-connected. Information is published and circulates with unprecedented speed, yielding striking images. For most people, however, social networking is a daily activity. This use of the mobile phone takes the many small periods of downtime during the day, and fills them with friends and news. It's a seductive distraction for individuals, and a potent problem for those seeking to hide the truth.

The Development of Mega-Cities

The tightest social networks are genetically coded into the behavior patterns of insects of the order *Hymenoptera*—ants, termites, and bees. Individually, an ant isn't particularly bright. It exhibits a total of perhaps twenty distinct behaviors. But social networks enable these near-automata to cooperate in much more complex ways. Through its social behaviors, an ant colony becomes a macro-organism. By working together, ants show swarm intelligence, and it's so effective that ants and termites now comprise about one-third of the biomass of all land animals.[2]

Many large-brained creatures also exhibit social network behaviors. These include lions, whales, elephants, wolves, and dogs. In Africa, for instance, dogs that hunt in packs succeed more than 80 percent of the time, far more often than individuals.[3]

Mankind's social patterns evolved in tribal settings over the course of millions of years. Tribes gave us power. Lone individuals or small family groups would have fallen easily to the big cats, but a tribe could better resist predators, hunt more effectively, gather berries and tubers, cope with disease, fend off other tribes, and employ the talents of individuals—athleticism, intelligence, weapon-making, wisdom—for the good of the group. Tribes worked because all of the members could interact. There was probably an optimal size for the tribe: not so small that a few deaths would endanger the group, and not so large that communication would become difficult.

Regardless, mankind eventually outgrew the tribal structure.

The first mega-city, with its 50,000 inhabitants, appeared at Uruk in early Mesopotamia, in the mid-fourth millennium B.C. It was there that humankind first experienced large-scale social structures with intertwined networks of trade, friendship, family, mutual interest, and political alliance. The dense concentration of people in cities made this network of networks possible for the first time in human history. They were unaided by technology, so physical proximity was needed. People had to see each other frequently—in the marketplaces, at weddings, at funerals, and at the city wells and washing sites—in order to maintain the social structures.

For several thousand years, social networking didn't appear to advance much. Patrician intrigues and popular insurrections in ancient Rome weren't so different from court intrigues and popular uprisings of the nineteenth century. Frequent interaction between groups of people is the key ingredient for creating a social system and no new technology appeared for this.

The advent of the telegraph, telephone, and railroad didn't really change the social dynamic much. While they increased the communication distances, they only connect two points at a time. They can't provide the rapid interaction between multiple points, so they can't duplicate the network effect that exists in a marketplace or town square. More recently, radio and television created a sense of global commonality—for better or worse—but the broadcasts were one-way. Lots of people could hear and see the same thing, but humans need to be part of a *conversation* in order to rally around an issue or establish a group consensus.

Our biological brains work by firing millions of signals through a dense constantly changing network of synapses that, in aggregate, galvanize to form a thought or memory. The denser the synapse networks and faster the signals, the smarter the brain and better the physical coordination of the body. The same tenet seems to be true of "social brains," too. The speed and density of social network interactions are the keys to forming group ideas and consensus, and taking action.

Thus, new social network technologies such as Facebook and Twitter provide a phenomenal amount of new network density—far more than any previous technologies, and certainly more than the town squares. When mobile computing is added to the equation, the density will double or *triple*, as more people acquire mobile computers. More importantly, mobile computing radically increases the speed of interaction by essentially eliminating the time lag between interactions. People can experience and participate in the public conversation all day long, with negligible delay.

This may be the beginning of a worldwide social consciousness. There is a vast network of nerve sensors that record and report everything that transpires in town squares, city centers, courtrooms, campuses, and byways across the world. It includes a complex array of synapses that processes those sensory inputs through postings, comments, "likes," and "shares." It involves a synapse network that is constantly evolving, creating new network connections and new pathways. The network can coordinate action among the moving limbs of society, creating flash mobs, protests, and book clubs.

The parallels between brain function and social network function are striking—enough so that it seems inevitable that the growing body of brain research will be necessary to understand the dynamics of social networking.

The Rise of the Social Networks

Facebook has become the busiest website on earth, and the closest thing yet to a master social network. In 2011, it reported 800 million users, while earlier networks like Friendster had mostly vanished from the United States, and MySpace was laying off half its staff.[4] Facebook outpaced them by understanding what makes a social network work, and by creating a technology that could scale to massive levels.

Friendster went online in March 2003 and became the first significant online social network. It was a closed system—each person could only see the profiles of friends. Its potential seemed vast, but its management focused on grand strategy rather than housekeeping. Its pages took as long as forty seconds to load, and it kicked out popular figures

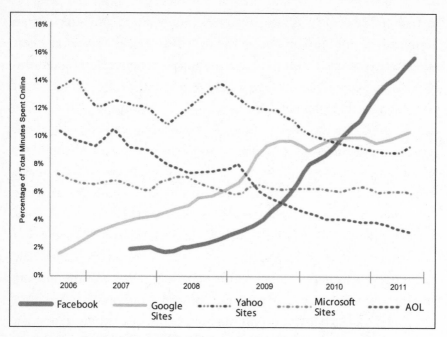

FIGURE 6.2 Facebook has been the total usage minute leader since 2011. It went from occupying 2 percent of total minutes spent online in 2007 to 16 percent in 2011.

Chart source: See Saylor. *Data Source:* Comscore, Feb. 2011 and Nielsen, State of the Media: Social Media Report Q3 2011 via Esteban Contreras, "The State of Social Media and Social Media Marketing in 2012."

like rock musicians for posting joke photos.[5] Members began departing, and a copycat site, MySpace, started in the fall of 2003.

MySpace courted the evicted musicians and their fans and offered an open system where anyone could see everyone's profile. Rupert Murdoch bought it in July 2005, and its unique visitors jumped 155 percent to 55.8 million a month over the next year.[6]

Facebook began in early 2004 and grew more slowly than My-Space, but soon shot past it with key innovations. For instance, in September 2006 Facebook introduced the newsfeed. This feature automatically sent your content entries—updates, comments, links, photos, "likes"—to all your friends, and gave you theirs. So you didn't have to visit everyone's pages to discover what was new. On Facebook, a "friend" could be a complete stranger with whom you've chosen to connect, and as you peered into his life, you got to know him (and vice versa). If you had 1,000 friends, they could all see the photos of your Bali vacation, and pass them on ("share").

MySpace waited eighteen months before copying the news feed.

Then in May 2007 Facebook opened its network to third-party developers who quickly created applications like the game *Farmville*. Facebook was giving its members something to do. The network became a playground, and this drew people further into it. MySpace lagged about a year behind, and before long Facebook was soaring past it.

Early social networks were largely hangouts, but Facebook is a social universe—a site for pastimes, albums, viral marketing, public announcements, autobiographies, and epitaphs. It's a broadcast medium for everything from morning yawns to killings caught on video.

As of 2011, Facebook dominates the United States and much of Europe. The monopolizing nature of information economics applies here. People don't want to check three social networks every day when one will do, and the one they choose will be whichever one appears to be the most popular. Facebook has conquered the United States, but Orkut—launched by Google in English-only—has come to dominate Brazil. QQ prevails in China, Cyworld in South Korea, Grono in Poland, Lunarstorm in Sweden, and Mixi in Japan. And Friendster remains popular in Indonesia and the Philippines.

Two other dominant social networking services fill different niches.

With 100 million users as of March 2011, LinkedIn is the professional person's Facebook. A combination social and business site launched in 2003, it helps professionals make connections, follow trends, find work, and simply chat. Offline, people obtain 50 to 60 percent of jobs through social contacts,[7] often through friends of friends, so a network that follows this pattern offers clear benefits. LinkedIn stores our resumes and business cards, and enables us to update them so the information stays fresh. The network is living and hot, and that's why organizations use it.

Many people thought Twitter was a joke at first. It began in mid-2006, created by a tiny San Francisco firm called Odeo, which faced lethal competition as a podcasting company and had to reinvent itself. The company came up with the idea for an ultra-simple way for people to broadcast their status, moment-by-moment, using SMS (short message service) text messaging on their phones. In internal testing, employees found it addictive.

Yet Odeo's board was skeptical. Who would use it, they asked? Just about everyone, it turned out, and that include celebrities, presidents, companies, protesters, and the guy next door. Most cell phones at the time could handle text messages only 140 characters in length, and so the famous Twitter limit arose, but it didn't matter. The number of Twitter users more than doubled in the second half of 2010, from 75 million to 190 million.[8] The seemingly humble site has become the bane of tyrants and the fastest source of news available on the planet.

The Evolving Social Phenomenon

Fueled by the immediacy and pervasiveness of mobile technology, social networks are gaining unforeseen powers and spawning unexpected new behaviors. In many ways, the social network is evolving into a sort of global organism—with each new subscriber, it expands its reach and knowledge base. We're not just connected to friends; we're connected to a collective intelligence with real-time global awareness. As the mobile-social phenomenon unfolds in the coming years we will see it apply in even more and perhaps unexpected ways.

The Personal Broadcast System: The original core use of social networking was as a broadcast system allowing each user to send personal news and content to his or her friends. Everyone feels they have something to say, and social networks provide us the podium to do that. We became our own small-scale broadcasters, with a friendly audience. Twitter took that idea much further with the idea that *anyone* could tune in to anyone's broadcast.

And it's fast. Facebook makes sharing almost instantaneous. You can snap a photo, post it, and within 60 seconds 3,000 people can see it in their news feed. It's almost telepathy. Using social networks this way lets us keep up with old schoolmates whom we haven't seen for years, even decades, as well as keep up on the latest activities of friends we see all the time.

The youth of today are spearheading the direction of social networking and mobile computing—hyper-connectedness. Always-on cell phones results in always-linked kids. When deprived of her mobile phone for 24 hours, one self-described "light" user said, "When I went to get it back, I felt connected again. I felt like I had gained back a part of myself, and I was back in business. I also felt a little more alive... like I had something in my possession that could occupy every spare moment of my life."[9]

Teens are connected to a degree their elders find difficult to imagine. By 2008, 42 percent of U.S. teens said they could text blindfolded, and 47 percent said their social life would end or be worsened without the phone.[10] By 2009, texting was more common than talking. Teens typically make or receive five phone calls a day, but half of the teens were sending fifty or more messages daily. Girls aged fourteen to seventeen were averaging 100 text messages a day, or six per waking hour.[11]

Proximity becomes irrelevant, and people are living wrapped in social contact that's afforded by mobile computing.

A Social Coordination System: The social network uses quickly extend beyond broadcast updates between friends—they help us coordinate our social activities. We see this already with the "check-in" and "event" features in social networks. These let people know where you've been in the past and present, and what events you will attend in the future. The frequency of broadcasting one's location to friends will undoubtedly increase through the use of mobile technology.

People will increasingly use social networks as real-time "social radar," to track the movements of friends. Suppose you have a thousand friends on Facebook and you can grade them on how much fun each one is. With a mobile-social app, on a Friday night, you could have a map that tells you the most fun nightspots based on how many and which of your friends are there. When you walk into a club, you could look down at your phone and see who's there, and gauge how much fun it might be.

People will use mobile-social applications to organize instant events. If you're going to a movie with some friends, you could determine which *other* friends are nearby, and invite them to join the group. You'd just hit a button, and zap an invitation to everyone within a fifteen-minute walking radius. Then they could accept or reject on the spot. With a single message, you would accomplish what used to require dozens of phone calls—not all of which would get through. Friends could accept your invitation based on who else was accepting. Coordination becomes instant and organic, rather than painstakingly sequential.

Mobile-social apps will help organize potentially longer-lived groupings, too. Officials have long urged carpooling, but without notable success, partly because it's difficult to accommodate constantly changing schedules caused by childcare needs or other unforeseen situations. A truly useful carpooling system would be highly adaptable to these changes and instant in its recalculation. A mobile-social app could let commuters offer rides at a specified street corner at a specified time to a specified destination. Then friends, or friends-of-friends, or even strangers could "bid" on the offer in real-time using their mobile phones. The first two or three bidders could be accepted by the offering driver—perhaps based on their Facebook profiles—and meet twenty minutes later at the designated spot. Everyone would know who to expect, and where, and when, with no surprises.

In one study, individuals extended this idea to include organizing groups to takes walks and ride the bus, just for the added safety and company inherent to a group.[12]

A News Filtering System: There is an avalanche of news available on-line. It's being created by commercial news agencies, online magazines, and innumerable bloggers and citizen photojournalists. It's impossible to scan even a small percentage, and it's very hard to determine what's important enough to be worth reading.

In the past, news agencies did this prioritization for us. Editors in newsrooms scanned all the news items proposed by their reporters, and decided on behalf of everyone what would be worth including in the newspapers, TV news reports, and magazines. Social networking is changing this.

In our social networks, each of us has a body of hundreds or thousands of trusted editors who scan different corners of the Internet and accumulate the important content for us, sharing links, posting editorials, sending synopses, and commenting on each others' editorials. It's a dynamically assembled news stream that should be highly relevant to each one of us, because the editors are our friends. Depending on your friends and their ability to judge, the news will be intelligent, prioritized, and pre-filtered.

The most important content in our social network bubbles to the top of our readers by virtue of how much each item is viewed, shared, commented upon, or re-tweeted. As we all participate more and more in this cooperative news service, it should get better reach and select the most relevant content for each of us. All of this will have great impact on traditional news services and the advertising dollars that flow into them.

Direct Connection with Consumers: Businesses and politicians also recognize the power of the socially powered news re-broadcast. Historically, organizations spent huge sums of money in mass advertising and public relations to get their messages out.

Today, with social media, they can send messages with virtually no cost of distribution. "Follow us on Twitter" and "Like us on Face-book" have become part of massive campaigns by many organizations to build audiences through a free social network channel. If a consumer "likes" an organization's Facebook Page, that consumer will receive that organization's messages through his newsfeed. As of this writing in 2011, many brands have massive Facebook audiences including the rapper Eminem with an audience of 50M Facebook fans, followed by Rihanna with 48M, Lady Gaga with 46M, Coca-Cola with 37M, and Starbucks with 26M.[13]

However, companies are quickly realizing that news is not enough to retain consumer interest, no matter how much people like their products. People want much deeper interaction that includes exclusive content, coupons, VIP-caliber services, special prices, and rewards programs. Even celebrities want to monetize the fan relationship by selling merchandise, tickets, and sponsorships. In essence, companies and celebrities need more than just a newsfeed to reach their fans and customers. They need a social-mobile application.

Why does this need to be a social-mobile application, rather than just a traditional website? The likelihood of a traditional web application becoming widely popular is slim. However, social networks hold a major key. They contain a wealth of personal data that can allow companies, for the first time, to produce highly personalized applications. Instead of offering everyone the same loyalty reward, using social network data, companies could offer each person a reward that is most pertinent to them. A music-lover, for instance, could be offered a free ticket to their favorite band, while a young parent could be offered a night out, while a student might be offered VIP access to an event that all her friends are planning to attend. Our "likes" and activities reside in social networks—most notably in Facebook—and can be accessed by Facebook-compatible applications. So it's finally possible for companies to achieve the much-sought-after one-to-one marketing relationship.

Mobile is important for this new breed of consumer because it makes the applications more available (24 x7) and because mobile can make the application *location-aware*. Imagine the impact of offering someone a free ice cream, just as they are passing by an ice cream shop. That might sound a little creepy at first, but who would be unhappy with a free ice cream cone on a hot day? And how much loyalty might such a simple gesture create for a company?

With mobile-social applications, retailers will increasingly offer bargains to passersby just to get them inside the store. Once there, customers might receive additional promotions, depending on the aisle they entered, the product they picked up, and their preferences as indicated by data in their Facebook profiles.

With mobile-social applications we'll also see faster liquidation of expiring inventories—*especially* inventories that will disappear at a set time. Seats on a bus and tickets to a concert or movie are time-sensitive inventories. If you don't sell the seat for the performance, it disappears as an asset. Here is how a mobile-social application might work at a multiplex cinema—it's 9:30, the mall closes at 10, and there's a movie starting in fifteen minutes. The theater has 250 empty seats and a pile of popcorn that's about to become garbage. With a location-aware mobile-social application, they can offer everyone in the mall who likes that genre of movie half-price seats and free popcorn, thereby monetizing that last chunk of capacity that otherwise would have gone to waste.

Or suppose you're looking for a dinner table for eight, fifteen minutes from now in your local area. You whip out your phone and it locates you. You enter "table for eight" at "8:00 p.m.," your mobile-social reservation application scans all of the open seating inventory, and offers you twenty-two alternatives organized by your favorite types of food and preferred price range. Perhaps it even notifies the restaurant and lets them bid on your business by offering a spot discount for something they've identified in your social profile. You might even get a greater discount by agreeing to post this restaurant as a check-in on your Facebook page, thereby implicitly promoting the restaurant to all of your friends. Everyone wins.

A Distributed Sensory System: Mobile computers provide the societal equivalent of a biological central nervous system that extends its nerve endings to all corners of the globe, sensing what is going on and reporting it to the rest of the "body."

The classic example is Twitter. You might be just one of 5 billion people in the world with a mobile phone, but if you see a violent crime in progress, you can type 138 characters, hit a button, and the tweet goes out to hundreds of your friends. If you had sent it to the *New York Times*, reporters might question your credibility, but chances are your friends do not. Within seconds, your hundreds of friends see the tweet and begin re-tweeting it, each blasting it out to hundreds of *their* friends. It goes from hundreds of people to *thousands*, in almost no time at all.

One nerve ending can sense an event and, in shockingly little time, large portions of the body known as "society" can begin assessing that event, to conclude whether action is warranted.

Add a compelling image and the sensory input is even more potent.

On June 20, 2009, a twenty-six-year-old Iranian woman named Neda Agha-Soltan was demonstrating peacefully in Tehran when the pro-government militia shot and killed her. A man with a mobile phone captured her bloody death on video, and it circled the world on Twitter, Facebook, and YouTube. CNN and PBS played it. It attracted global attention and became a rallying point for those who oppose the Iranian government.

However, a tweet may be a rumor or an outright lie, and if enough people find it credible and interesting, it becomes a *misinformation* wave.

On November 4, 2010, a tweet falsely claimed that a Qantas Airways flight A380 had crashed in Indonesia, and the firm's share price plummeted in value until its executives contacted the media. In the Massachusetts Senate special election of January 2010, an activist group sent a last-minute smear-tweet of the Democratic candidate, which reached 60,000 people but did not affect the outcome. Yet the tactic is cheap, easy to use, and potentially effective. As Filippo Menczer of Indiana University—who investigated the Massachusetts incident—noted, "From the point of view of someone running for office, it would be crazy *not* to use this system."[14]

A Societal Coordination System: More than a hundred people converge on a statue of Queen Victoria at Sydney, Australia's town hall, to bark like dogs and hiss like cats. Then they quickly disperse.

In San Francisco, a group forms suddenly, spins dervish-like in the street for ten minutes, and then vanishes.

In Rome, faux bibliophiles appear in a bookstore and request nonexistent books. They break into applause before disappearing.

Flash mobs are a form of urban pranksterism—apparently spontaneous but organized in advance by mobile phone. The first occurred in Manhattan in June 2003, when a group 100-strong suddenly appeared in a Macy's department store in Manhattan asking for a "love rug" they could "play on." Then they vanished. Bill Wasik, a senior editor at *Harper's*, had organized the phenomenon, but even he was surprised as flash mobs spread across the globe.

From Sydney to Tokyo to Toronto, mobs have held zombie walks, silent raves, subway parties, and pillow fights. They've worshipped a dinosaur and performed synchronized swimming in a public fountain. In 2004 the term "flash mob" entered the *Oxford English Dictionary*.

Flash mobs seem like great feats of organization. It takes time and energy to coordinate a large group, and in the old world no one would bother, just to accomplish a prank that would last only a few moments. But today, with one-to-many communication, the task becomes relatively easy. Planners don't have to call a hundred people—they just post invitations, or send them to mobile phones. They can use the technique for purposes beyond fun, including soccer hooliganism and riots.

And uprisings.

On January 18, 2011, two weeks after ousted president Ben Ali had fled Tunisia, a twenty-five-year-old computer company employee named Asmaa Mahfouz was distributing flyers in Cairo's Tahrir Square. Security officers pushed her around, so she went home and taped an impassioned video of herself urging citizens to come to the square on January 25—the holiday known as National Police Day. She then uploaded it to Facebook. Within days it got 78,000 hits, people emailed it further, and posted it on YouTube.

On January 25, as she had implored, 25,000 demonstrators came to Tahrir Square—and to squares all across the country. The crowd occupied the Square and refused to leave. By January 28, the number of protesters had swelled to hundreds of thousands.

So on February 2, the government of Hosni Mubarak sent in plain-clothes security officers. Violence is usually pretty effective at breaking up a demonstration, but in this case tweets captured the experience as it happened:

waelkhairy88 Oh God. The sound of gunfire! Machine guns echo can be heard everywhere. God help us all. *05:01:00*

monasosh 2 of my friends confirm another one is shot through the head, dead. My friend called me crying #Jan25 this is awful, something has 2 b done *05:29:33*

waelkhairy88 Updates: 13 injured and 4 dead as a result of a few minutes of gunfire. *05:34:04*

waelkhairy88 Gunfire came from above the 6th of October bridge . . . protestors went up there and arrested them *05:38:51*

mosaaberizing A final wave with few armed thugs carrying machine guns taking place now. Some more martyrs. #Tahrir *05:46:14*

waelkhairy88 The tanks are moving. Two dead men are being dragged from bridge after being shot dead. *05:47:21*

mosaaberizing We're fighting on the bridge now. They don't exceed 100. We need to hold on for 30 more minutes. #Tahrir *06:06:20*

mosaaberizing Sunrise in Cairo. Blood spilled in Tahrir more noticed now. All over the place. #Jan25 *06:39:10*[15]

This crowd held fast, and the tweets may have been crucial to morale. Moreover, anyone could receive and re-tweet them, so these riveting messages spread worldwide. Mubarak resigned a few days later, and many think this episode undid him.

Dictatorship does not occur easily in tribes. Tribal leaders often gain sway through influence, not force. In tribes, everyone knows everyone else and it's easy to organize a counter-movement when there is 100 percent transparency and access to all of the participants. But large-scale civilization changed the dynamic. With cities, centralized wealth, and organized government, the ruling class gained far greater control. A few people could rule a multitude by dividing and isolating the opposition. A political machine can have swift, clear communication top to bottom, along with well-understood roles, incentives, and weapons.

That's why social networks are so pernicious to authoritarian governments. A dictator can blackmail, arrest, or kill ten ringleaders. But it's hard for him to blackmail, arrest, and kill 25,000 people without risking serious, perhaps international backlash. Through social networking, he's attacked by a swarm of hornets, instead of a rhinoceros.

Information today moves, not just with unprecedented speed, but with great precision. It can bypass every news organization on earth. If the images are compelling, the information stream takes on a life of its own, being rebroadcast virally from network to network. And if the message warrants it, the societal brain will make the decision to act.

Harnessing the Wisdom of Crowds: We've always reached out to our friends and elders to seek wisdom and advice, whether the topic is relationships, child rearing, which television to buy, or where to go on vacation. Social networks extend this ability from a small circle to hundreds or thousands of friends, and potentially to millions of people participating in the same opinion-aggregating applications.

The most obvious idea is to simply use the social network to post questions to one's friend network, and read the various answers that come back. This is good for entirely free-form questions that can have an infinite range of answers: "What should I do about my mother-in-law?" "What are some of the best strategies for saving for college?" "What are some fun things to do this weekend?"

A more scientific and statistical approach would be a simple "wisdom-of-the-masses" application that lets us distribute "voting polls," for which the range of answers would be chosen from a fixed, multiple-choice list. We could ask: "Who will win the election?" "Which of these cars should I buy?" "Which of these wines is best with fish?" Voting polls enable us to tally the frequency of each answer, so we can see what the majority thinks, and how divergent opinion is. If the wisdom-of-the-masses application uses the social data of the participants, it can organize the answers more intelligently by telling us what the men recommend, or what the women recommend, or what people who are most like me recommend.

The crowd may have broad experience that will provide valuable insights, even if no one member has explicit expertise. Or we can fish the crowd for someone who has explicit expertise on a topic. Either way, making this application social lets us distribute the question to people whom we know and trust, and making it mobile enables us to get the answers faster.

Another use of our wisdom-of-the-masses application would be for "numeric polls" where the question asks for a number as its answer. For example: "What will the Dow Jones index hit tomorrow?" "How long should it take to drive from Springfield to Shelbyville?" "On a scale of 1 to 10, how much can we trust this candidate?"

There is a body of psycho-mathematical research that shows that large crowds of independent people can guess numeric values and predict outcomes with great accuracy, when their responses are averaged. This was shown embarrassingly in 1907 when Englishman Francis Galton tried to prove that masses of uneducated people could not make good judgments. He asked more than 800 people at a county fair to guess the weight of an ox. The guesses varied widely, both high and low, with many being patently ridiculous.

However, much to Galton's chagrin, the median of all the answers was strikingly close to the actual weight and was a much better estimate than had been offered by a handful of experts he had employed to guess. In the theory of wisdom of the crowds, even though no one might be an expert, each person brings a little bit of information to the problem. Overestimates balance underestimates, often resulting in a reliable solution.

With social networks, wisdom can be gathered without explicitly asking questions or even requiring participants to respond. Facebook, in particular, possesses a lot of demographic information that each of us has willingly entered—name, location, age, education, marital status, number of children, etc. What's not so widely known is that Facebook also contains an immense amount of information about what you enjoy, based on the pages you have "liked," locations you have "checked in," and events for which you have RSVPed.

From this psychographic information, a social-analytic application could analyze your friends to determine what type of music they like best, what artists they follow, and what they enjoy in terms of books, movies, hobbies, restaurants, sports, athletes, and celebrities. If you wanted to keep up on music trends, you could analyze which bands have accumulated the most new "likes" from among your friends. You could find out which clubs have become most popular based on check-ins.

As people purchase more and more items through social commerce sites, you will be able to find out which products are most popular among your friends without having to poll them.

Taking the concept even further, there will emerge more and more social-mobile applications whose purpose is simply to gather distributed information. With Trapster, founded by Pete Tenereillo in late 2007, drivers with mobile phones spot speed traps, traffic-light cameras, hidden police cars, and other hazards, and report them to a clearinghouse which aggregates all of the input. It's a wisdom cooperative, and a descendant of CB radio, once used by long-haul truckers for similar purposes.

As of January 2011, Trapster had nearly 10 million members.[16]

Other applications already have wisdom built in as an ancillary feature, and more will follow. For instance, OpenTable captures reviews of every restaurant, so if a majority of diners post negatives, you'll find out before you go there. You can access assessments of doctors, attorneys, sporting goods stores, and nurseries. And as you become aware of the reputations attached to others, you may pay more attention to your own.

You might wake up one day and realize that you have an eBay-like performance rating attached to you.

The Universal Identity System: On April 15, 2009, Boston police found the body of a young masseuse named Julissa Brisman in the Copley Marriott hotel in Boston. The aspiring model had advertised under "Erotic Services" on Craigslist, and one of her clients had shot her.

Police got a good look at him. He was a young, clean-cut man who walked calmly down the hall after the murder, texting on his BlackBerry and apparently unconcerned about the surveillance cameras. They were able to obtain an email he had sent, tracked down its Internet provider address, and identified his computer. They arrested and jailed the man—a former medical student named Philip Markoff. Authorities planned to charge him with attacking two other women he had met on Craigslist, but he committed suicide before he could stand trial.

The press dubbed Markoff the "Craigslist Killer," but he was not the first.

Almost a month before the Brisman killing, New York radio reporter George Weber was stabbed fifty times, allegedly by a troubled youth he had met through a Craigslist post offering $60 for rough sex. Two weeks later, a teenager named Michael Anderson was sentenced to life in prison for the October 2007 shooting death of a twenty-four-year-old,

Minnesota preacher's daughter who had replied to his fake ad for a baby-sitter. Anderson, too, was given the "Craigslist Killer" label.

Contacts through this site have also led to an array of thefts and other crimes.[17] On Craigslist, you know nothing about the persons with whom you are dealing, except what they might have told you on email or over the phone.

Bill Warner, a Florida private investigator, says more than half his business involves running background checks on potential Internet dates. Married men or stalkers—the typical threats—can join a site using a false name, a throwaway cell, and a misleading email address from a free service like Hotmail.

"There are a lot of people out there who get jazzed up by disguising themselves," he says.[18]

Yet there's no reason our social network profiles couldn't be used in a positive way—as an identification system. If we had easy access to the social profiles for everyone with whom we deal, we could approach them with confidence. More than just his or her correct name, we would know the town in which that person lives, where they went to school, and the name of their spouse. If we had friends in common, those linkages would help ensure that our dealings would be ethical.

And safe.

It's common to see fake driver's licenses—all that's needed is a forger and a photograph. But it's much harder to concoct a fake persona on Facebook, where 99.35 percent of Facebook users provide their real names.[19] A false persona on Facebook would be very difficult to establish and maintain with hundreds of faked friends, hundreds of faked tagged photos, and multiyear timelines of activity.

An identity system based on Facebook could be even more secure if linked to other networks. For instance, you need a link to a bank to use the PayPal network. That linkage makes PayPal an authenticated network. Both PayPal and your bank agree that you are who you say you are, and they have a history of bank transactions to back up that claim. Thus, if you see a PayPal link on the Facebook page, the likelihood is greater that the person is legitimate.

You can also link a Facebook profile to a passport. You can take a photo of your passport, attach it to your Facebook page, and have the Justice Department verify it. All of a sudden you would have a "certified account" on Facebook, and showing someone your Facebook profile is like showing your passport.

Mobile-social identify applications would enable you to authenticate people you meet instantly. At a party, you can check the legitimacy of a new acquaintance prior to accepting a date. You might not yet know the person, but you could identify three friends of a friend, skim their photos and profiles, and *then* strike up a conversation. At a conference, attendees could wear virtual name tags offering their profiles, photos, and friend lists.

All manner of businesses could benefit from such identity systems. If someone gets into a taxicab and robs the driver, the driver has no recourse. But before accepting a passenger, the driver might require the person to scan in a Facebook profile or use their digital cash account for payment. In either case, who would dare commit a crime when he's been placed at the crime scene?

Parents might ask for the Facebook profile of a new babysitter, to know a lot more about who is watching over their child. A housewife might ask for the Facebook profile of the handyman who's coming to fix the sink, to feel much better about opening her house to him. If the parties of a Craigslist transaction exchanged Facebook profiles prior to meeting, then neither party could perpetrate a crime and expect to get away. There would be a full digital record of the visit.

The only reason we don't have instant identification systems available to everyone, is that up until now they haven't been cost-effective. Today they are. And they'll knit us all together into a much tighter, more secure, more credentialed society. Over time it will become harder and harder to live outside of such a society.

And there will be fewer places to hide.

CHAPTER 7

MEDICINE

The New Landscape of Global Health Care

The Paper Swamp: Medical Records – Telemedicine – The Hospital as a Network – Portals for the Deaf and Blind – Global Medical Care – Third World Prometheus

"Because it's there," George Mallory told reporters, when they asked why he wanted to climb Mt. Everest. Lively and charismatic, Mallory set out in 1924 with Andrew Irvine on a celebrated attempt to scale the world's highest peak. As they approached the summit on the North-East Ridge, a cloud obscured them.

They never returned. Their fate remained the biggest mystery in mountaineering until 1999, when a party discovered Mallory's frozen corpse. He had slid down the face of the mountain, and something had punctured his forehead, leaving a wound that looked as if it came from an ice axe. But why did he fall? The best guess is that he'd grown dopey from oxygen deprivation.

If he attempted Everest today, he might not have died at all. With mobile technology, doctors all over the world can monitor the blood oxygen of Everest climbers, and they could have sent him back to camp.

In the United States, health care consumed 17 percent of the U.S. gross domestic product in 2009. That's a huge amount, higher than that of any other developed nation, and more than triple the 5 percent figure released in 1960.[1] Americans spend $6,102 per capita every year on health care, about twice as much as Canada ($3,165) and three times that of the United Kingdom ($2,083). Yet the average U.S. life expectancy remains below that of most European nations.[2] The health care system is rife with waste and snarled in regulation. Pricing is complex and often opaque, and physician quality is hard to gauge. The payer–payee relationship is indirect, and the intricacy of the system adds needless costs.

Mobile technology can streamline many of the processes and could save hundreds of billions of dollars. Doctors will move to mobile because it's there, and because it can remove inefficiency and friction from their industry.

The Paper Swamp: Medical Records

Modern medicine is like a city with gleaming towers and large, primitive slums. Remarkable advances like stem cell therapy have enabled the blind to see again by restoring the optic nerve, yet medical *records* remains a backwater. As early as 1991 the Institute of Medicine concluded that electronic medical records were essential,[3] yet by 2010 just 20 percent of doctors in the U.S. had them.[4]

Even the trucking industry is more digitized.

In medicine, paper has bred the same costs and handicaps we've seen elsewhere, and more. Information is stuck to the page, so it doesn't move easily. Physicians are forced to manage records in file cabinets and warehouses, wasting time, labor, and real estate. Records can't be kept close at hand, they can't be accessed quickly, and they can't be searched efficiently. They can't be viewed in advance for emergency patients. They can't be duplicated without a copier. And records, especially handwritten ones, lend themselves to errors. Patients suffer, attorneys prosper, and premiums rise.

FIGURE 7.1 Paper medical records take up valuable space in a doctor's office and finding, writing, and organizing each file wastes valuable time.

The medical record is a creature of accretion. In the early nineteenth century, most physicians were rural general practitioners (GP). For instance, George Huntington, who published the classic description of Huntington's chorea in 1872, was an obscure, music-loving doctor in New York State. He and other practitioners kept medical records as simple personal files, often covering a patient's entire lifetime. As knowledge expanded in the late nineteenth century, specialties arose. By 1930, one quarter of physicians were specialists, and about two-thirds are today.[5] Because a patient now sees more than one doctor, far more professionals have to read each record and add to it. In addition, medical technology has made tremendous strides, and there is a great deal more information available for each patient. So the files have grown.

Thus, today's medical record contains information on vital signs, medications, prescriptions, allergies, referrals, lab results (including

X-rays), insurance records, payment accounts, and more. It has a covey of authors, is largely handwritten, and serves medical, legal, administrative, and billing purposes, as well as acting as an archive for future research and the teaching of medical students. For outpatient visits, a physician can spend up to 38 percent of his time locating information and writing observations.[6] As one medical director at a major hospital said, "The charts are so thick with everyone documenting, documenting, documenting, you can't make your way through them."[7]

Ultimately, everyone pays.

Electronic medical records have been available for years, but they've faced varied forms of resistance. Doctors liked paper because they could jot notes while looking directly at the patient; computer use disrupts eye contact. Electronic records have been costly. The software itself has often been complicated and difficult to learn. Even so, if the U.S. medical world had gone electronic in the late '90s, by one estimate we could have saved thousands of lives and billions of dollars.[8] With everyone connected, doctors could easily access information twenty-four hours a day. They would rarely repeat tests, because the results would be available to all, and emergency responders could know an injured person's entire medical history before even reaching him. Centralized information isn't foolproof, but it's simpler, faster, more accessible, and ultimately less expensive.

The Obama administration required doctors and hospitals to go digital by 2014 or—if they're in the Medicare system—face penalties starting the following year. The 2009 stimulus bill pumped billions of dollars in incentives into this effort, while the 2010 health care law set up more programs to encourage the use and study of digital dossiers. With electronic records on file, the medical system will be poised to utilize mobile technology, a move that will further enhance the medical chart system in many ways.

For instance, the handwritten chart will become obsolete, and electronic records on a desktop computer already can replace it. One study found that the amount of time it takes an employee to pull and replace a chart costs $5, and since electronic records reduced chart pulls by 600 per year per physician, those alone could save $3,000 per physician.[9] The mobile chart is more convenient, as well. The doctor can always have a tablet computer at hand, as can the nurses and technicians. It can be written on while the doctor maintains eye contact with the patient, just as if it was a paper chart, and handwriting recognition can convert the scrawls into text that can be searched and indexed. Patient interviews can be recorded, and speech recognition software can be used to transcribe the interview. NFC communication will make it possible to automatically upload medical data from the bedside diagnostic equipment. Test results can be beamed to the mobile record so doctors have them as soon as they are available.

France has used a form of mobile records since 1998. Every person over eighteen carries a small plastic card called the *carte vitale*. Its magnetic strip holds that person's entire medical history: every doctor visit, prescription, referral, test, procedure, and more, along with the doctor's fee for each visit and the amounts the insurance and patient paid. When a new patient arrives, the doctor slides the *carte vitale* through a reader and this history appears on the computer screen. The physician types in recommendation such as a drug prescription, it goes at once onto the *carte vitale*, and the patient can take it to a testing service or pharmacy. The doctor also knows at once which insurance company to bill and completes the billing with a few keystrokes. Information is encrypted; so French officials say it's private and safe.

As a result, large file cabinets have vanished from French doctors' offices, and hospitals have 67 percent fewer staff assigned to handle paperwork.[10]

Mobile record keeping will make the *carte vitale* obsolete, too. The *carte* requires specialized readers not readily available everywhere a doctor might be. Doctors and nurses move about constantly. They make rounds. They go from one ward to another, to different departments, and to conference rooms and operating theaters. As long as records can only be read at the desktop, they will often be unavailable.

Tablets will alleviate this problem, granting swift access to lab and examination findings, diagnoses during a hospital stay, all relevant documents, and much more.

Even at a bedside, doctors will use mobile technology to access medical literature and drug inventories as they speak with the patient. When they make diagnoses or give instructions during rounds, they can enter them in the record on the spot. The mobile record will allow them to watch for potential drug interactions, checking dosages against weight and blood pressure readings, and note the time previous medications were administered.

Perhaps most intriguing, mobile technology will enable medical records to become multimedia. Instead of describing patients, doctors can photograph or film them. Other physicians can see a wound or rash, observing with detail far beyond the capacity of written descriptions. They can see the patient's behavior, hear his or her voice, and view progress over time. They can justify interventions better, teach better, and better fulfill every *raison d'etre* of the medical record.

The mobilization of health care can't occur without physician buy-in, but mobile technology is a natural for them. Indeed, it's already happening. Almost all doctors own cell phones, and in 2010 more than half of them regularly used smart mobile devices for everyday treatments, up from 25 percent in 2004.[11] They inhabit a mobile world already.

Prescriptions will become electronic. As in France, you can carry the prescription with you on the device, rather than a separate slip of paper. Electronic prescriptions eliminate the possibility of losing them and allow the patient to change pharmacies without hassle. Prescriptions themselves can grow more sophisticated, displaying photos and perhaps using video to explain side effects. We can know our medications better and make fewer mistakes with them. A mobile application should even remind you that it's time to take your medicine, and ask you specifically about side effects—feeding the answers back to your doctor, if necessary.

With mobile devices, patients can also keep their own medical records. To take a simple example, more than two-thirds of U.S. adults are overweight or obese,[12] and excess fat breeds slow body wreckage. But some apps let you track your calorie count right at the table. Web sites can deliver customized meal recommendations, readable in the supermarket or anywhere you shop. In addition, dieters can snap a photo of each meal, upload it, and get a professional's opinions and suggestions. In essence, the dietician is looking over your shoulder.

Telemedicine

Around the year 400 B.C., Hippocrates urged doctors to place an ear on the patient's chest to check for a heartbeat. The practice fell into disuse, a long time passed, and in 1816 French physician Rene Laennec sought to revive it. But he faced a problem: Women balked at the intimate contact. So one day he rolled sheets of paper into a tube and touched the end to a patient's chest. To his surprise, the beats were *clearer*. The tube amplified the sound. After much testing, he developed a stethoscope that was simply a wooden tube. Laennec would go on to make many discoveries about the lungs and, using his invention, would learn at forty-five that he was dying of tuberculosis.

But his tube changed medicine. Previously, doctors had relied mainly on subjective accounts: "I felt hot. Then I started coughing and threw up." In isolation, these descriptions were limited and often misleading. The stethoscope made assessments objective. The practitioner could measure the true workings of the body and learn facts no patient could articulate. The device became a symbol of medicine itself, and appropriately, since it was the first great breakthrough in diagnostic technology.

Thousands more would follow.

Now these instruments are migrating to mobile technology. For instance, the early app iStethoscope picked up the heartbeat when an iPhone was held to the chest, and soon people downloaded millions of copies. It was a very basic app, and a person needed skill to find the body's four key listening points, but it had virtues the physical stethoscope lacked. It showed patients a waveform of the heartbeat and kept a digital record of it, so they could compare patterns over time. And it had another advantage: You could send the waveform anywhere. A million doctors could be at the other end of that stethoscope.

Telemedicine isn't new. Medieval Europe stoked bonfires to relay warnings of bubonic plague. In the 1960s NASA monitored astronaut heart rates, respiration, blood pressure, and temperature from outer space,[13] and pioneered satellite-based communications to areas beyond the reach of radio waves, such as aircraft and inaccessible mountains and valleys.[14] If humans land on Mars, we'll hear each of their heartbeats with about four minutes lag time.

What *is* new is telemedicine that is ubiquitous and affordable. By 2004, even cows were having sensors placed under their skins at a cost of around $100 each, to monitor for signs of mad cow disease in the heartbeat, body temperature, and other functions.[15] Such sensors can be life or death for people with chronic illnesses like heart disease, emphysema, stroke, cancer, and diabetes. These are the world's leading killers, responsible for 60 percent of all deaths,[16] and they afflict more than 125 million Americans.

Ten percent of patients generate some 70 percent of health care costs, and most have chronic illness.[17] Even partial successes in this realm will save hundreds of billions of dollars.

In developed nations, about half of the chronically ill—almost 65 million people—don't take their medicine regularly, because they forget, worry about side effects, can't pay for it, or simply ignore the doctor.[18] This failure is suicidal, in a probabilistic sense, and every year some 90,000 Americans die prematurely because they disregard instructions for high blood pressure alone.[19] It also wastes the time of medical staff and drives up expenses.[20] The bill for it overall is $100 billion a year.[21]

Mobile devices excel at reminders. The notices can be automatic text messages, and each vial of medicine can come attached to a scale, so that if its weight doesn't drop regularly, the bottle can warn both doctor and patient.[22] Mobile devices can't force a pill down an unwilling throat, but through the use of video they can bring in a clinician to discuss the matter. The doctor can assess the situation remotely, and by reducing unnecessary outpatient visits, prescriptions, and admissions, such monitoring could cut annual drug costs by 15 percent.[23]

Vital signs such as heart rhythm, temperature, blood pressure, blood oxygen, and blood sugar levels all act as the body's early warning system. Abnormalities imply risk, yet we usually gauge them sporadically, at the doctor's office, and run the serious risk of missing the ongoing pattern. But with body sensors that send signals to our mobiles, we can track them constantly, study daily charts, and check our vital signs as easily as we open email.

There are 74 million people in this country suffering from hypertension,[24] and when they all have sensors, we will manage blood pressure better. Mobile telemetry is helping diabetics maintain their blood glucose, and it can alert the 7 million undiagnosed U.S. diabetics to their disease.[25] It can also warn the 79 million pre-diabetics and help them prevent the condition. Asthmatics can gauge the pollen, smog, and other triggers outside, and see alerts when levels rise. We can track our sleep patterns with electroencephalograms and review minute-by-minute charts the next morning. Such ability will open a hidden realm to inspection, but it will also map irregular sleep patterns and make them easier to treat.

Physicians can monitor us from afar. Already, wristband sensors track vital signs and alert doctors to potential problems. If you are one of the 2 million Americans at risk of congestive heart failure, you can wear a pressure gauge in your pulmonary artery that quickly notifies doctors of trouble and may save your life.[26] A heart patient can also wear an electrocardiogram (ECG) monitor as a belt clip or a neck pendant. It sends a waveform of the heartbeat to the mobile device, and if the software detects a danger sign and clinicians validate it, the physician is informed at once. Each year in the United States we lose over 400,000 people to sudden cardiac death, and when any doctor on earth can observe your live ECG, that number will plummet.

In emergencies, mobile technology can bring expert care faster. For instance, if you're watching a baseball game at the stadium and feel sudden chest pain, you can see and talk to a doctor on the phone, transmit your vital signs, and get advice in an instant. If you need to go to the hospital, the specialist can accompany you all the way, for mobile technology places the physician into the ambulance. If you're unconscious, each paramedic can still send live video to the hospital, and doctors can study close-up images, hear heartbeats and breathing, and view real-time video of the response to treatment.

Time is the enemy in a heart attack, and many studies show that a quick response saves lives. It preserves cardiac tissue, and dramatically improves the quality of life afterward.

Mobile technology will shorten hospital stays, since sensors can monitor a patient's progress at home. The average hospital charges $2,129 per day,[27] so the economies are plain, and most people prefer their homes to the hospital.

Demographers forecast that the global population over 65 years old will reach 761 million by 2025, twice the number we had in 1990.[28] The old will outnumber the young for the first time in history. As humanity grays, we can lighten the cost and burden on long-term care facilities like nursing homes. With mobile telemetry, we can keep the vital signs of the elderly under constant surveillance, so their homes are much safer for them. Sensors on everything from refrigerators to doorknobs will enable remote care-giving conducted by family and support groups. They will also reduce loneliness.

In the end, such advances will kill the symbol of medicine itself. A doctor with a stethoscope will inspire as much confidence as one with Laennec's wooden tube. Already, a portable ultrasound device the size of a cell phone lets doctor and patient visualize the heart muscle, valves, rhythm, and blood flow, and send the information to a mobile device.

"Why would I listen to 'lub dub,'" said cardiologist Eric Topol, chief academic officer for Scripps Health in San Diego, "when I can see everything?"[29]

The Hospital as a Network

The hospital was originally inseparable from the temple, since disease was thought to come from the gods, and was remorseless in its effect. Hence early hospitals like the Hôtel-Dieu of Lyon (est. 542) and the Hôtel-Dieu of Paris (est. 660) paid more attention to the patient's soul than his body. Yet medicine became objective, expanded, and grew more specialized through the nineteenth and twentieth centuries. As it did, doctors began to work in teams, and hospitals became large institutions with growing staffs of health care professionals.

For instance, the X-ray analyst came to be present in many health facilities, but he might read just a few images each day. Soon, however, there won't be a need for him to be on-site at all. Already an iPad can store and display an X-ray so detailed that it can be used to diagnose the patient. Medical imaging had long been done strictly in-house, but tele-radiology has spread quickly, and by 2007 44 percent of radiology practices themselves were using it for analysis in the off-hours.[30] Radiology has many sub-specialties, and mobile technology makes it easier to send X-rays to an expert who can give a better diagnosis, no matter where that person is located.

In the intensive care unit (ICU), critical-care specialists or "intensivists" are experts at detecting post-operative complications and minor anomalies that might lead to death. Mortality rates are 30 percent to 40 percent lower in hospitals where they provide round-the-clock care.[31] Yet there is a dearth of these experts. Only about a third of patients in the ICU today receive care from an intensivist, and the U.S. Department of Health and Human Services projects that the shortage will continue for thirty years. But there don't need to be intensivists at every facility—not when they can treat people remotely.[32]

An intensivist in a hospital might handle ten beds, but a single doctor and four nurses based in a remote command center can oversee seventy-five to one hundred patients. In the year 2010, such systems cost $4 million to $5 million on average to install, and could cost $2 million annually to staff and maintain. But hospitals say they soon pay for themselves in reduced costs, lower mortality rates, and shorter lengths of stay in the ICU. The University of Massachusetts Medical Center alone saved $5,000 per case because remote intensivists ordered treatment for patients sooner than they otherwise would have received it.[33]

The big hospital is like the big box store—Target, Wal-Mart, and their brethren. Each one is designed to have at least one of every kind of specialist on hand. But to the extent that heart scans, blood telemetry, X-rays, and other medical information can move, and patients can see specialists on video, co-location no longer matters so much.

So a much more efficient architecture can be created. The hospital will become much more of a network, and much less of a big box. There will be a blossoming of clinics, because a neighborhood center will be able to take tests and send results to a remote specialist.

Rural areas have suffered from a lack of hospitals due to their low population density. So a low-density rural area needs a low-density approach, like a clinic. And clinics can work in other places, too, like the school and the mall. With network-hospitals, there will be an opportunity to bring medicine much closer to the consumer and at a cheaper price, as well.

Portals for the Deaf and Blind

Mobile technology can be a great gift to deaf and blind persons, since it will bring them sense data wherever they go. For instance, a simple cell phone currently is more or less useless to the 25 million deaf people on earth, and at five or even twenty-five words per minute, texting is inconvenient and slow. But mobile computing technology gives them portable video, so they can speak anywhere employing sign language, facial expressions, and lip reading. We convey 120 to 200 words per minute with both sign and spoken language,[34] so mobile phones can let the deaf talk more naturally to others, even from far away.

Mobile devices—such as bracelets, belts, or watches tied to the Internet—will provide visual and physical alerts (such as vibrations) to notify deaf people of incoming calls. These same devices can act as audio sensors and signal deaf person of a home break-in, the wails of a baby, or the fall of an unseen object.[35]

As information becomes more and more visual, deaf students may prefer to attend classes via the tablet. Every class will have the option of being subtitled, and well-known professors will teach students scattered across a nation. Indeed, as speech recognition grows more sophisticated, subtitling will be available for all conversations.

There are 1.3 million legally blind people in the United States.[36] Independence and safety are obvious concerns for them, and the mobile device can help.

With an earpiece connected to a device in a person's pocket, and using GPS and a digital compass linked to a database, that person can walk through the world with pre-existing knowledge. The mobile phone can query the GPS chip, learn the user's location, and translate it into words spoken into the ear. For example, "Robinson's Hardware is twenty feet front of you," or, "The picnic table is on your left." The mobile device can be a tour guide, providing information, history, and context about the world around us. And the users can talk back to it, ask it questions, and garner more precise information about the environment through which they travel.

Mobile technology offers the blind or visually impaired (BVI) affordability, flexibility, and effectiveness. Frameworks like MoBraille connect phones to Wi-Fi-enabled Braille displays.[37] Hence mobile devices can send information to BVI persons in Braille or speech. (In surveys, blind participants testing mobile devices prefer Braille to speech output.[38]) And some apps can use crowd-sourcing to enrich visual or auditory maps with crucial, real-time information—for instance, that a bus stop has moved. Ultimately, devices like RFID chips positioned throughout the world may help the BVI better navigate their homes, their friends' homes, and the larger environment.

Global Medical Care

People have searched abroad for better health almost as long and diligently as they have searched for treasure. Mediterranean folk in search of cures trekked to the Epidauria festival seeking Asklepios, the Greek deity of healing. For hundreds of years, mud baths, hot springs, spas, and sanitaria have drawn the disease-ridden to all points of the compass, from the Adirondacks to the Alps.

Ever-mounting health care costs, low airfares, and the global spread of medical technology drive today's medical tourism. The savings are their own argument. Dental work in Mexico costs a fifth of that north of the border.[39] A liver transplant that runs $300,000 in Chicago would be $90,000 in Taipei.[40] A heart-valve replacement in the United States would cost at least $200,000, but $10,000 in India— including round-trip airfare and a little sightseeing.[41] A knee replaced in the United States costs five times what it would in Thailand.[42] And the bargains get even better for Lasik eye and cosmetic surgery, for which well-trained and equipped specialists in Latin America have years of experience.

Moreover, U.S. health insurance often doesn't cover orthopedic surgery such as knee and hip replacement, or it may impose significant restrictions on the choice of facility, surgeon, and prosthetics.

U.S. general practitioners average a salary of $186,000 per year, and specialists $340,000,[43] while Indian GPs earn $5,260.[44] So the trend here is obvious.

Physicians overseas often see patients faster. Waits in the United States are longer than those in Germany, France, Sweden, Denmark, and most of the developing world, though shorter than in Britain and Canada.[45] Instead of idling until slots open up in the doctor's schedule, as would often be the case in the U.S., patients go to the head of the line.

According to one estimate, by 2018 nearly 16 million Americans a year could be seeking cheaper knee and hip replacements, nose jobs, prostate and shoulder surgery, and even heart bypasses. The lost revenue to U.S. hospitals and providers could be billions per year.[46]

Yet medical tourism will become increasingly unnecessary as mobile technology brings distant services to local areas.

In medicine, video communication is more than forty years old. Doctors forged the first interactive video link in 1964 between the Nebraska Psychiatric Institute in Omaha and the Norfolk State Hospital, 112 miles away. The first complete video telemedicine system appeared in 1967 and tied Boston's Logan Airport to Massachusetts General Hospital. It not only transmitted video images, but X-rays, medical records, and lab data.[47]

Interpersonal dynamics remain intact with video. Psychiatrists have used it with few problems. It hasn't reduced the rapport between patient and doctor, nor the perception of emotional subtleties. Even the paranoid patient has rarely felt extra anxiety. Indeed, psychiatrists find that over the course of the treatment, the awareness of the medium vanishes, and both sides feel as if they are talking directly to each other, as if through a window.[48]

You'd talk to a doctor in Bangalore the same way.

Suppose you were at a mall, and suddenly felt nauseous and disoriented. You might find a quiet bench, open your mobile device, and dial a physician in India.

"The world is swaying," you'd say. "I feel woozy. I may vomit."

He'd download your medical history, if he didn't already have it on hand, then use body sensors in your mobile device to examine your vital signs. This could be as simple as using the mobile camera and an app that can detect your heart rate and respiration. As you watched, he'd review your history, read your vitals, and interview you, taping you all the while for the records. The doctor then would download a prescription to your mobile, which you could fill at a nearby pharmacy. The whole consultation might cost $10, there would be no waiting, and you might prevent a more serious and costly episode.

We can place medical kiosks or mobile clinics in public schools, and make it cost-effective to check children regularly. We can put them in senior centers and in malls. We can make public health widely *public*.

In the Third World, medical service centers will arise. Instead of calling one doctor, you might call a network that's staffed with perhaps a thousand. If a general practitioner didn't have the answer, a specialist could come online. If necessary, a follow-up referral to a live doctor could ensue.

And of course such centers could do much more. There's no reason why the specialist analyzing your X-rays or monitoring an intensive care unit has to be where you are. He, too, can be in Bangalore, and many teleradiologists already are. "Remote" means anywhere.

Then, if a patient needs surgery, he might travel to India for it, or have it done in the United States. But, as surgical robots become common, the patient might stay here and have a surgeon perform the operation remotely.

As mobile medicine puts downward pressure on prices in the United States, pent-up demand will emerge. More people will see doctors and receive advice and treatment sooner. Early detection and preventive care will further decrease costs. We will live longer, healthier, and better.

Third World Prometheus

Whatever the health care level may be in the United States, it's Plato's "Ideal State" compared to the situation in the developing world. Here in the West, more than two-thirds of people live past seventy, and most die of chronic diseases. In middle-income countries, nearly half the people live to seventy, and chronic diseases are also the major killers—though tuberculosis and traffic accidents also play a role.[49]

But in low-income countries, less than a quarter of people reach seventy, and children under fourteen make up more than a third of the deaths.[50] UNICEF and the World Health Organization (WHO) estimate that malnutrition ends of lives of 6 million children under five years old, every year.[51] In comparison, cholera kills just over 100,000 people annually. More than one billion people lack access to clean water, and death commonly arises from complications during pregnancy and childbirth.[52] Infectious diseases are major killers, including lung ailments, digestive diseases, HIV/AIDS, and tuberculosis. Tuberculosis alone took 1.7 million lives in 2009, and malaria 850,000.[53]

Sub-Saharan Africa is the world's hardest case, a teeming bestiary of illnesses. The HIV virus thrives there as nowhere else, infecting 46 million people and killing 18 million as of 2011.[54] Nearly one-third of the children are underweight,[55] and 43 percent lack safe and accessible drinking water.[56] Two-thirds lack adequate sanitation.[57] Illnesses scarcely known in the West thrive in the region, and are known as NTDs (Neglected Tropical Diseases). These affect more than one billion people worldwide—a seventh of the globe—and every year they steal 57 million "disability-adjusted life years," a measure that reflects both death and poor health.[58] More than 90 percent of this loss stems from seven disorders: hookworm, whip-

worm, ascariasis (roundworm), lymphatic filariasis (the cause of ele-
phantiasis), onchocerciasis (river blindness), schistosomiasis (bil-
harzia), and trachoma.[59]

All are controllable.

Societies in sub-Saharan Africa have always been isolated and frag-
mented. Just a fifth of its people have access to electricity,[60] and only
29 percent of the roads are paved.[61] Want to mail a letter? Postal systems scarcely exist, and people hire runners to take messages on scraps
of paper across cities. There are three landline telephones for every
hundred people, and a person can wait two to ten years to get one.[62]
The telephone companies themselves are notoriously inefficient. The
governments themselves are usually their largest customers, and since
they can ignore the bills for years, the companies starve financially.[63]

In this part of the world, mobile technology is a Prometheus. In
fact, it bestows a far greater gift than fire. It confers eyes, ears, voice,
reach, and *knowledge*. It can bring forlorn, dusty villages the greatest
communications medium in history.

Citizens everywhere in the developing world have snapped up the
phones, and two-thirds of the world's 5.3 billion mobile phone users
live in emerging markets.[64] In most of Africa, the per capita income is
less than $1,000 a year,[65] and mobile accounts can be relatively expen-
sive. Yet by 2011, more than 40 percent of the one billion residents of
Africa already possessed mobile phones—up from 0.4 percent in
1998.[66] Many Africans are semi-literate or illiterate, but with the
multi-touch interface they can join the online world.

And the benefits start with health.

There is no medical system in sub-Saharan Africa—not as we understand it. Approximately 80 percent of its people have never seen a doctor, and rely on "traditional healers." In rural Mali, for instance, some
people think evil spirits bring on trachoma.[67] And a patient may not
even bother with a doctor, if convinced that supernatural forces cause
this blinding disease.

So on a fundamental level, mobile phones will educate patients. By accessing websites, Africans will learn about ailments—their symptoms, causes, and treatment. They can request the proper help, discuss their conditions more productively with providers, and help diagnoses improve. They and their families will suffer less.

Mobile technology can also educate health care workers, to boost their number and caliber. Since most providers in Africa aren't doctors, the Internet represents an enormous opportunity for them, and they can train at a fraction of the cost of brick-and-mortar schools. Mobile devices can also inform them of the cost, quality, and suppliers of medicines and other commodities. It can foster market competition, with its typically lower costs and better performance.

Mobile technology can provide lifelines. Rwanda has one of the world's worst records for maternal death in childbirth. Only 5 percent of Rwandans have electricity, and roads are generally unpaved, so most women give birth at home. Many die of bleeding or infection, both of which are easily prevented. The Rwandan government has given *hundreds* of mobile phones to community health care workers so they can track pregnant women in their areas, send monthly checkup reminders, and dispatch ambulances if problems arise or women go into labor.[68]

Small medical test centers can spread across the continent in areas devoid of Western medicine. People who have never had blood, saliva, or urine tests will get them. Diagnoses will come from top specialists, rather than traditional healers. This will lead to the demise, not just of the mediocre, but of the downright awful.

In the nation of South Africa, Project Masiluleke has sent millions of text messages encouraging people to take tests and receive treatment for HIV/AIDS.[69] In Zambia, two strategies have proven effective in reducing HIV/AIDS in Africa. One is a nationwide fidelity campaign,[70] and the other is circumcision. A Zambian group uses mobile technology to schedule operations, answer questions, and remind patients of post-operative care.

"When I think about the biggest impacts, I think about patient re-minders," said Bill Gates, co-founder of the health-oriented Bill & Melinda Gates Foundation. "Malaria and TB are going to be the first things where you say, 'Wow, without this mobile application, all these people would have died.'"[71]

Tuberculosis is the world's most lethal infectious disease. The drugs to treat it can cause nausea and heartburn, and patients must take four to twelve pills a day for six to twenty-four months. So some neglect them, especially when the symptoms start improving after a month or two. Yet those people give TB to others, and strengthen drug resistance in the bacillus. In doing so, they are courting the grave.

So monitoring is crucial to scale back this disease, but in the poor re-gions of the globe where it thrives, direct observation is rarely possible. Mobile technology can keep watch over them. For instance, the pillbox itself can send a signal to a server whenever the patient opens it, and if no signal arrives, staff can call that person with a reminder. In some tri-als in South Africa, this approach has freed nurses to handle fifty to sixty patients, where before they were dealing with just about ten.

A determined fool can thwart such a monitoring system, but he can't thwart biosensors in the skin—which doctors have also used suc-cessfully.[72] And with video, a nurse can actually watch a patient take his pills, just as if she were there in person. Such strategies will spread throughout the developing world, to areas where there is no feasible alternative.

Dr. John Snow made the world's most famous disease map when—during an 1854 cholera outbreak in London—he plotted the deaths in Soho. His map showed that they clustered around a single pump and proved his theory that cholera spread through infected water.

In modern society, atlases of disease abound in the West, yet we still lack precise maps of malaria and most NTDs. Scientists only fin-ished the first detailed map of sleeping sickness in 2010, after a ten-year effort that dug down to the village level. As a result, anyone can

THE MOBILE WAVE 165

track the disease on a mobile device and see the contour lines of the threat. When disease maps are produced for illnesses that have cheap, effective drug cures—like trachoma and hookworm—health workers will be better equipped to plan programs that will administer the drugs where they are most needed.

For diseases that are more difficult to combat, the arrival of such maps will be even more pivotal.

Can we exterminate malaria? The parasite arose in reptile intestines at least 60 million years ago, far earlier than the 35-million-year-old mosquito. It long daunted Westerners and they didn't start colonizing Africa until after 1850, when the new drug quinine reduced mortality by four-fifths.[73]

In some parts of Africa, malaria still infects nearly all children by the time they reach the age of two, causing weakness, waves of high fever, and often death.[74] Official programs to eradicate the pestilence go back to 1955, and as a result the global malaria map has shrunk. Yet it is far from complete, and 3 billion people remain at risk. Mobile technology can accelerate this challenging task. Specialists need to learn of new cases as soon as possible, so they can intervene and slow the spread. Communication in Africa has been so wretched that administrators have relied on irregular assessments that don't supply enough data. In addition to the lack of information from the field, key obstacles have included weak infrastructures, shortfalls in funding, lack of political will, and administrative inflexibility.[75]

Trials using mobile phones have begun, however, and have already proven effective—as well as inexpensive—in tracking the disease. Though few expect a zero parasite count any time soon, the devices may provide a vision of success, and begin to ease the political situation, as well.[76]

Measles is another target for extinction. Campaigns have wiped it out in the Americas and East Asia, and elsewhere the number of cases has taken a nosedive. Worldwide, the measles deaths of children under five fell from 733,000 in 2000 to 118,000 in 2008.[77] Death rates have risen since, however, because of a fraudulent scholarly paper appearing in *The Lancet*, purporting to show that the measles shot abetted autism. This led to a campaign by distressed activists with little sense of science, and inoculations dropped off.

Measles is highly contagious, but with mobile technology the medical community can spot outbreaks swiftly and inoculate kids in nearby areas. Vaccinations cost less than $1 US per child, so there's no overwhelming economic barrier.[78]

Schistosomiasis is an infection by a fluke that can live in the body for up to twenty years, continuing to lay eggs, and one medical professional described the sweeping effects in vivid detail. "The victims' bellies swell to accommodate a damaged, swollen liver; energy is replaced by apathy . . . industry languishes. Worse, a vicious circle sets in: less food is grown by these tired people; they come to suffer increasingly from malnutrition, and lose hope."[79] In Leyte province in the Philippines, each schistosomiasis victim was losing 45.4 days per year to the ailment. Treatment with praziquantel cut that number to four, and the economy gained 41 more productive days annually per person treated.[80]

Trachoma and onchocerciasis cause blindness, while hookworm and malaria induce fatigue, and all have a distinct societal impact. Trachoma alone subtracts $2.9 to $5.3 billion annually from world productivity. When rates of onchocerciasis reach 10 percent, farmers often abandon fertile land for hardscrabble. After officials bring it under control and the farmers come back, the economic rate of return improves by up to 18 percent.[81] Hookworms can lay 30,000 eggs each day in the body of a child, and in some areas this has been found to cut that individual's future earning capacity by up to 43 percent.[82]

When it comes to controlling disease, the economic upside has been apparent since at least the early twentieth century. After Greece eradicated malaria in 1974, for instance, its economy soared. And in a classic virtuous cycle, stronger economies lead to better health. Healthier people are more productive, and they spend less on medical costs. Capital grows. Clean water becomes available. Governments drain malarial swamps. Indoor plumbing arrives. Diet improves. More money can flow to education and infrastructure. Outsiders become more apt to invest, and tourists to arrive and spend. Health care itself gets better.

Mobile devices will spin this cycle faster. In the United States, life expectancy grew from forty-seven in 1900 to sixty-eight in 1950, and the increase stemmed almost entirely from the conquest of infectious diseases.[83] Mobile technology will pull the developing world down this path, too. As it does, their people will live much longer and better lives.

CHAPTER 8

EDUCATION

Remaking Education for Everyone

Education Matters – The New Textbook – Active Learning and
Virtual Worlds – The Digital School Office – Nobel Laureates at the
Lectern – One-to-One Learning – Breaking the Curve of College
Costs – Global Education

When Singapore became independent in 1965, it was a seedy, low-slung, malarial island-city of 1.9 million.[1] Pale shop houses lined the streets and sold groceries and gewgaws, tropical scents wafted through the night, and dark mildew stained the walls. The fabled Raffles Hotel had grown run-down, and guests sweltered without air conditioning. Singapore had a fine deep-water port on the Strait of Malacca, yet the nation sat on just 217 square miles of land and had to import all of its water, food, and energy.

Its per capita income was one fifth that of the United States.[2]

Today it is a throng of gleaming skyscrapers, a spectacle of light in the evening. Double-decker buses rumble past shopping arcades housing Tiffany, Ferragamo, and Cartier; the subway is swift; and no one complains of a lack of air conditioning. The average citizen lives eighty-one years.[3] The gross domestic product of this dot on the map exceeds that of Finland and Israel, and it has the fifth highest GDP per capita in the world.

The United States is eleventh.[4]

What was the secret? Much of its success stemmed from the combination of a free market and a disciplined society engineered by Lee Kuan Yew. He and his advisors steered the economy from Third World to First, astutely if sometimes strictly. For instance, tourists can visit Singapore's casino complex for free, but the croupier's rake is for foreigners, so Singapore citizens must pay $100 to get in.[5] However, Lee Kuan Yew saw from the start that education was a *sine qua non*. The average Singaporean in 1960 had three years of schooling—less than the average Paraguayan.[6] By 2010 almost half the citizens had high school degrees, and more than 90 percent had a tenth grade education.[7] Some completed polytechnic or vocational training, but all were employable after school. A 2007 study from McKinsey & Co. rated Singapore's education system one of the best on earth.[8]

Nothing has forecast the growth of a city better over the last century than its education level. After Gutenberg, the towns that adopted printing early grew more prosperous, probably because their residents knew more. The impact can be surprisingly strong. Among nations, just one extra average year of education correlates with a 25.8 percent increase in output per capita.[9]

Formal education is almost as old as writing. The first known schools appeared in Sumer around 2500 B.C., in Shuruppak, home of Noah of the Mesopotamian flood legend. They arose to teach cuneiform to scribes, and as the clay libraries grew, schools offered other subjects, too.

For millennia, schools served the prosperous, since most roles in societies were fairly simple and didn't requiring formal education. By the nineteenth century in the United States, though, schools were spreading everywhere, albeit slowly. Most pupils attended small rural schools where they learned to read and write. But as that century progressed and turned into the twentieth, society changed fundamentally. Cities grew, technology proliferated, and markets spouted wealth. The economy needed far more educated people. In 1900 just 6 percent of

U.S. youths had obtained high school degrees, but by 1940 50 percent of them had done so.[10]

Education is the portal to the intricacies of modern life. The brain builds representations of the world, and schools add depth to those constructs. The better the representations, the better the individual can negotiate life. Biologically, these representations are simply connections between brain cells. Broadly speaking, human beings have two kinds of memory: short-term and long-term. Short-term memory lets us handle the immediate environment, and then deletes the minutiae—which is almost everything. But its contents can turn into long-term memory, or *knowledge*, when neurons receive intense or repeated stimuli. They then grow extra synapses—extra connections to other cells—and the hippocampus integrates the memory with the rest of the brain.

If you remember where you were when you heard Osama bin Laden was killed, it's because of the flurry of new, stable synapses that were created. So schools cultivate representations—from the spelling of "something" in the earliest years to the traits of microRNA in graduate school—and heighten our skill level. Thus, the better the university we attend, the higher our economic value. (U.S. college graduates earn an average 80 percent more per year than non-graduates.[11])

Learning quickens the information flow. Each well-educated person tends to know a variety of things that are different from the ones known by his peers. Less-educated people, however, tend to know many of the same things as one another. Thus, better-educated people have more information to pass along—in working groups, discussions, and social networks. By doing so, they enhance their skills, and their income.

Everyone benefits. If the average number of college graduates in a metropolitan area rises by 10 percent, the average salary increases 7.7 percent, no matter what each individual's education level may be.[12]

Education breeds invention, which creates wealth. Technological innovation usually requires a social web such as Silicon Valley, where ideas can build upon other ideas. The tablet computer is a perfect example—it was a confluence of technologies going back to the mainframe and before. Without an educated social web, innovation languishes.

Education is the underpinning of our knowledge economies, yet the U.S. system faces trouble today. Even accounting for inflation, the price tag for education from kindergarten through grade twelve has quadrupled over the past thirty years.[13] We spent $476.8 billion in 2007,[14] but it still seems too little, and fiscal angst abounds. Though California paid $96 billion to educate children in 2010, 30 percent of its students were attending a school that was in financial distress. Thirteen districts were on the brink of insolvency, and many more were facing it.[15]

What are we getting for this expense?

In a 2011 evaluation of education in sixty-five nations, the United States spent more money per student than any other except Luxembourg, yet it ranked fourteenth in reading, seventeenth in science, and twenty-fifth in math. U.S. scores roughly matched those of Estonia and Poland, but each of those two paid 60 percent less to teach each student through age fifteen.[16] In many of our failing schools, as few as one in five students can do grade-level math or English. In some, it's as few as one in twenty-five.[17] Our low educational achievement costs the nation as much as $175 billion per year.[18]

Meanwhile, the endemic problems remain. For instance, schools have long had trouble finding good teachers for courses such as physics. As the world changes ever more swiftly, schools will face further difficulty hiring teachers for newly important subjects such as Chinese. Textbooks have long cost too much, wear out too soon, and quickly become out of date.

And many students still look blankly at the fabulous trove of knowledge available to them. In one 2009 survey, two-thirds of students were bored every day, and one-sixth were bored in every class.[19] It's pretty hard to learn if you don't care. Our brains allow short-term memories to vanish in order to avoid store-housing trivia, and if a student judges the actions of Alexander Hamilton to be trivia, his brain will save its resources for other information—perhaps the deeds of professional athletes.

All of this is poised to change. We've lived with the original Sumerian education model for 4,500 years, and it's ready to evolve.

The New Textbook

"Your future is all used up," Marlene Dietrich said to Orson Welles in *A Touch of Evil*, and he wound up dead in a ditch. Printed textbooks face the same prospect. They have even less of a future than books in general.

Physical books will go upscale, but printed textbooks will enter oblivion.

Remember that three-year-old in St. Tropez who was working on his music? Imagine what he'll think when he enters kindergarten and finds teachers telling him to read static books, and he is forced to learn to write in cursive script on paper. The world will soon be full of preschoolers who have grown up with the magic screen, and in school we'll try to yank them back to the last century.

The tablet computer embodies education and all its involvements. It's the best didactic technology ever invented. It can give you every textbook. It can deliver all the basic reference manuals, indeed, the whole library. It can send you a test, let you take it, grade it and return it at once, then certify the results. It can provide video resources, including lectures, and help you do group projects across the globe.

In 2010, U.S. school districts were paying more than $8 billion per year for textbooks.[20] A single book could cost $100 or more. In just one state—North Carolina—the four-year tab for each high school student was about $1,020. That's $122.8 million spent statewide for each four-year cohort.[21] A tablet costs far less than $1,020—about $600 per student (including service), and it can also last four years. Moreover, with Moore's Law and the bargaining power of a state like North Carolina, the price could be brought down. And as tablets become ubiquitous, schools will have to buy fewer and fewer for children who lack them.

What about the cost of information?

Much lies in the public domain. Roslyn High on Long Island in New York launched one of the first iPad test programs in 2010–2011. Officials realized that 60 percent of their books in English class, such as *The Adventures of Huckleberry Finn*, were available for free as downloads, and that they could easily update outmoded works, such as biology texts that omitted new topics like tissue engineering.[22]

In time entire textbooks will be free, as well. Several U.S. companies offer free online texts, and make their money by selling add-ons like print-on-demand and audio materials. One nonprofit organization offers "flexbooks," which are free, customizable textbooks that meet state standards. They can come in varied levels for different tiers of students, and anyone can download them. By 2009 Virginia had adopted a physics flexbook, and while he was in office California Governor Arnold Schwarzenegger announced a plan to replace science and math texts with them.

Across the Pacific, South Korea declared pen and paper "obsolete" and launched a program to give all students digital texts by 2013.[23]

Digital textbooks are multidimensionally richer than print. Movies once required a film projector and a pull-down screen. Now they're right in the book, and instead of just reading about Franklin Delano Roosevelt, students can watch a clip of him speaking. They can look up terms like "palea" in an instant, and build vocabulary in the best way known—by seeing (and hearing) words in context. They can view and rotate simulations of, say, the insulin molecule. They can answer questions in chapters, see errors at once, and keep going till they get all of them right. They can take multiple-choice tests and view scores instantly, and teachers can learn which topics a class hasn't understood. They can practice foreign languages at home, with endless examples— including audio of the proper pronunciations. They can join forums, link to other sources, and search for data. And with special subjects like music, they can hear at once the material the instructor is addressing.

Educators have a habit of adhering to tradition. The Akkadians taught students in Sumerian long after the language had died. The Romans taught in Greek till the empire was gasping, and the seventeenth-century Europeans taught in Latin. Science had to battle its way into university curricula in the late nineteenth century.

Today's schools will cling to print texts, too, as educators strive to unlearn their old habits. But freeing the face of the vast economic realities, and the undeniable benefits of mobile technology, we'll rip this $8 billion per year out of the education system. That's the first step toward letting it breathe again.

Active Learning and Virtual Worlds

Over the course of the iPad test program at Roslyn High, students revealed that the devices enabled them to be better organized. Instead of carrying binders or multiple notebooks, they had all their notes in one place. So it was easier both to take the notes, and to study them later.

Likewise, students in a pilot program at Gibbon Fairfax Winthrop High School in Virginia also said they were better organized. More than 85 percent were doing their homework on iPads, and students found the devices most useful for science, English, and calculus.[24] In math, students could easily write equations and draw graphs. They could also email their homework to teachers, and get it back by the same route.

Other forms of pedagogy are emerging. One English teacher at Roslyn High started giving students small assignments in the evening before class, to get them thinking ahead of time. In the small district in Canby, Oregon, elementary school students used a two-minute math app to refresh their skills as they moved from recess back to class. Such spaced repetition builds synapses and fortifies memory, and the time spent on the app was the equivalent of increasing the school year by six days. [25]

The best learning is active, and there's a wealth of evidence indicating that active learning gets students more interested and boosts recall.[26] Students absorb more by working—or playing—with information, than by passively assimilating it.[27] A 2010 study in *Nature Neuroscience* found that we learn better when we have more control over the material and concluded that memory is "an active process that is intrinsically linked to behavior."[28] The brain is a builder, not a sponge, and it forges meaningful linkages around information. That's why learning a new word in context improves recall, and why teaching a subject is such a good way to master it.

Tablets are made for active learning. The screen reacts to human commands. In a text, students can read about Victoria Falls and see a photo of it. In a digital text, they can view the water plunging over the cliff, hear its roar, and move around it at will, penetrating the crocodile-ridden waters of the Zambezi and the geological strata under the falls. They can *experience* it.

When two or more students work together, combining their skills, it's known as collaborative learning, a highly active process that has proved to increase exam scores from the fiftieth to the seventieth percentile, and cut the dropout rate in technical fields by 22 percent.[29] Collaboration *per se* is a key skill to learn, since teamwork is essential in many industries, and workers today collaborate with people from other cultures across the globe. Mobile technology is reinventing collaborative learning, enabling students worldwide to meet in web-based forums to create databases, wikis, image boards, and other resources. Essentially these are social networks, and often the students are more adept at them than their teachers.

Game playing is also active learning. Elementary schools have long used single-player games like Reader Rabbit and Math Blaster, but we are starting to see collaborative games in alternative realities. NASAs *Moonbase Alpha* is a multiplayer online role-playing game in which kids become astronauts and embark on realistic missions to learn about space exploration and destinations like the moon and Mars.[30] Synthetic Worlds Initiative at Indiana University is developing a virtual game called *Arden* in which students will enter the world of William Shakespeare and learn about and the playwright, his works, and the history, culture, and economics of his era.[31]

Tablets also give access to virtual worlds that are not games. The premier example is *Second Life*, a social universe so richly immersive that some have sacrificed notable portions of their real lives in order to participate. Colleges teach courses within the confines of the *Second Life* virtual world, discussion groups meet, and students can attend from anywhere.

Edusim is a virtual learning world where students handle 3D digital objects, usually through the use of a multi-touch screen.[32] They can play with detailed 3D renderings of dinosaur fossils and study the movement of birds mid-flight, viewing the image from every angle. And with the "Open Wonderland" open-source toolkit, other organizations can easily create their own robust, secure, 3D virtual worlds.[33] *Arden* and *Edusim* reinforce real-world instruction, but their basic strengths lie in group learning.

Tags make the world into a showroom, as we've discussed, but they also make it a museum. With tagged and augmented reality, students can wave the phone and discover the age and history of buildings on a college campus. Entering a lab, novices can learn what every piece of equipment does and how to handle it properly. The environment itself becomes a teacher.

Tourists are students, in a sense. In Paris, for instance, guides describe the history of the Sorbonne and the Eiffel Tower. But a tagged Paris will tell you far more, in whatever language you speak. Students on field trips are like tourists, and tags will make their learning more active as they hunt down and collect information about sights such as sun bears in a zoo.

The Digital School Office

Tablets make the classroom more cost-effective, but their impact on the educational system goes further. Like hospitals, schools still depend on paper for administrative needs and record keeping. Mobile devices will eliminate its shortcomings and slash its costs.

For instance, teachers can use a mobile device to take roll call in elementary school. Kids can check in and out of every period, administrators can track the children through every period at school, and the information enters the district's computer instantly and automatically. Process steps disappear.

It's easy to lose records on paper. In 2011 an eight-year study showed that 21 percent of student records were missing from several California schools.[34] Yet access to such information not only helps teachers and counselors work with students, but also determines how the state allots funding.

The Roslyn Union Free School District began its transition to mobile administration by giving tablets to each of the seven members of its school board. The devices cost a total of $4,200, but the district estimated it would bank over $7,000 the first year and $11,000 each in the second and third, for a total of $29,000 over three years. The board would save eighty reams of paper each year, as well as time and labor spent in filing, copying, discarding, and more.[35]

And that's just the seven-member board. The district reported that it created 17 million paper copies in 2008–2009, but in the pilot iPad program that number was close to zero.

The district enrolls about 3,400 students, and we can use its results to extrapolate what the effects would be on a national scale. There were 57,523,000 total K–12 students enrolled in the United States in 2009.[36]

Based on the Roslyn numbers, that's 276 billion paper copies nationwide every year.

Schools are paying for a vast river of paper they don't need.

Nobel Laureates at the Lectern

There are more than 14,000 high schools in this country, and there's at least one calculus teacher for each. Test them and you could find the best, the Jaime Escalante of the movie *Stand and Deliver*. What is a Jaime Escalante worth? Teachers trooped into his classroom trying to copy his methods, but few succeeded. Yet if Jaime Escalante had used the tablet as his podium, he might have been able to teach 500,000 students at a time, offering each of them the best calculus education available.

This concept can be taken even further, to usher in experts. Stephen Hawking could lecture on cosmology and Ron Howard on filmmaking. Bill Clinton could teach politics. The open-source movement at colleges has taken us partway there already. An entire array of universities makes educational course materials freely available to students. For instance, through MIT's OpenCourseWare program, top college professors teach calculus on Internet video. Similarly, at Yale well-known professors such as economist Robert Shiller and historian Donald Kagan offer free video courses.

The template already exists. And the concept can be *intuitive*. For example Bill Clinton would go to a studio and record the answers to a large set of questions, such as:

"How does a president affect a bill moving through Congress?"

"What did you do on your first day in office?"

"What's the most important quality a president needs to possess?"

He would also record answers to follow-up questions, and all of the material would be tagged and cross-referenced. Thus, if a student asked any variation on one of the queries, the computer would fashion a response so that it sounded just as if the President himself were replying. Just as the Google search bar anticipates words the user is likely to type, programmers can anticipate phrasings. "Chat-bots" are becoming more and more convincing, and the technology behind them can make the material more realistic—even conversational. One chat-bot is the Artificial Linguistic Internet Computer Entity, or ALICE. Its broad knowledge base and interpretation software make ALICE pretty convincing.[37] So as long as the questions remain within the speaker's knowledge space, the exchange between student and expert could take the form of a convincing dialogue.

This process resembles the famous model from antiquity: Alexander the Great taking tutelage from the best-known scholar of his time, Aristotle. Soon all students will have their Aristotles.

So How Much Would We Save?

As of 2009, there were about 3.1 million K–12 teachers in public schools, with an average base salary of $49,030, for a total payroll of $152 billion each year.[38] The open-source instruction from MIT and Yale, on the other hand, is free. And even if an expert like Hawking or Clinton charged a million dollars for recording the material that would be used, that would be a one-time price for material usable for years by millions in educational systems worldwide. We'd shed billions in needless cost and solve endemic problems.

Long-time teacher shortages—for example, in science—would vanish. If material from world-class educators became readily available, school districts wouldn't be forced to settle for instructors who would be, at best, adequate. Students would be exposed to first-class minds, gaining a clearer, more dynamic understanding of each subject. The programs could be updated and kept current with findings about, say, dark matter or the Higgs boson. Public schools would stay on top of the changing world.

Instructors could be chosen based on their unique skills in delivering information to target audiences. Some might specialize in teaching the gifted, others the average, and still others the learning-challenged. Escalante, for instance, excelled with the underprivileged.

School districts will also get the teachers who can best master the new medium. This won't necessarily mean the best lecturer. Schools will want the best "tablet teachers," the persons most able to grasp and marshal the full dimensions of mobile education to engage and educate students. Teaching criteria themselves will change in a fundamental way.

Classroom teachers won't disappear, but there will be fewer of them, and they will have new roles. With a certain expertise in hardware and software, many may act more as guides, support, and facilitators for learning groups. They will also be needed to proctor exams. For instance, as one Japanese teenager was taking a college entrance exam in 2011, he snapped photos of the math and English questions, uploaded them to the Internet, got the answers in minutes, and entered them on his test. Authorities caught him,[39] but it's unclear how many cheaters they *don't* catch. Teachers will always be needed to help keep testing honest.

So we slash that $476.8 billion a year[40] in costs. And we get a dividend of billions per year that we can use to rescue flailing school districts, buy tablets for every student, enhance educational facilities, cut dropout rates, and make education more effective overall.

One-to-One Learning

The inertia of our educational institutions is immense, but the pressure of change forced upon them through mobile technology innovations is also great. In the next twenty years we will reach an inflection point.

The quality of learning you will be able to obtain for yourself will exceed the quality of the learning you would receive from a teacher in the classroom. This inversion will be very threatening to the teachers unions, yet it's nothing more than common sense.

Mobile technology will make it possible to import instruction. If a doctor can treat a patient, face-to-face, from Bangalore, a teacher can also teach this way. At $50 an hour, most families can't afford private tutoring, but at $3 or $5 an hour it will be hard to resist. The pupil can learn calculus from a graduate student in New Delhi, economics from a professor in Chengdu, and physics from a Nobel Prize–winner in Sweden. Already tutors from India and China are teaching children in Manhattan via video links, and the trend will accelerate.

Many public schools are ill-disciplined and even scary. Almost half of new teachers find other work within five years, and when instructors are bailing out of an environment, you can't expect parents to want their children there. Many private and charter schools are filled with kids whose parents mainly wanted them out of public school. Homeschooling serves the same end.

If you are a parent with multiple children to teach at home, you can use a tablet to acquire flexbooks, download lectures, watch videos, and administer tests. The tablet is both the schoolroom and the teacher's aide. Homeschooling families can choose from thousands of examples of online learning software, course curricula, audio and video lectures, and texts. Thousands of downloadable lectures already exist on the web, created by professors for their students. Some of these were done years ago and float on long-forgotten websites, like messages in bottles. A homeschooler can seek out these resources, assess their appropriateness, and put them to good use.

Homeschooling also allows parents to keep pace with brain research in a way schools have not. For instance, right after a lecture, students can recall 70 percent of the information from the first ten minutes, but just 20 percent of it from the last ten.[41] So fifty-minute periods are inefficient, and parents can break sessions into ten-minute chunks to help children consolidate their memories. Similarly, most education systems still start teaching foreign languages in high school. Yet the evidence indicates that the younger you begin, the better you learn. If an English child between one and three is exposed to Chinese language, for the rest of his or her life her left hemisphere will process Chinese grammar just as it does with English. But wait till ages four to six, and both hemispheres will handle the grammar. This is a clumsier arrangement.[42]

We live in a global economy, and with every year the knowledge of Chinese and other languages becomes more valuable.

Breaking the Curve of College Costs

I grew up in a middle-class family. My father was an enlisted man in the Air Force and he rose to become a chief master sergeant, the highest rank possible for a person without a college degree. When I graduated from high school, our family's net assets were $3,000—and I went to the Massachusetts Institute of Technology (MIT), where the tuition was $10,000 a year. I couldn't possibly have attended without substantial grants. I could have gone to a community college or a state university, but without Air Force financial aid, MIT would have been out of reach. So I indentured myself to the Air Force to attend MIT.

America is a rich country, and it was even richer then.

Today we're impoverishing ourselves to maintain our system of higher education. Tuition has risen almost twice as fast as inflation since the late 1950s,[43] by an average of 7.25 percent per year (compared to inflation's 4.35 percent).[44] So families now struggle to pay for college, and by the time students graduate, they are deeply in debt.

Why has higher education become so costly, even as society needs it more and more?

There are several explanations. One cites the competitive zeal of universities. Each school strives for greater prestige than its rivals, so each spends more and more on staffing and equipment, in an endless race. Another highlights the tendency of students to ignore the cost of a top-flight institution, since they feel they can pay off the loans in a few years, and batten off the brand forever. And the brand can have great value. If a Princeton grad and University of California grad were vying for the same job, whom would you bet on?

Since 1950, the prices of dental, medical, and legal services have grown almost in parallel with higher education. All involve personal services that require many years of formal education.

Intriguingly, the cost of hiring a stock broker rose at about the same pace until the early 1980s, when they stopped rising and began to plunge. The reason is clear: People discovered that they could buy stocks from their desktops and no longer needed all those well-trained professionals. Online brokerage systems were providing better service at far less cost.[45]

In higher education we'll see the same transformation.

Princeton costs about $55,000 each year for tuition, room and board, and books, adding up to $220,000 over four years.[46] For this sum students receive quality facilities, food, living quarters, peer connections—and the Ivy League brand.

Yet with resources accessed via mobile technology, a college student can get a certified education that is essentially as good as Princeton's for a few thousand dollars, total. Tablets can open the door to remote, effective, less costly adjunct professors. An instructor from Nairobi can appear online and teach the same course curriculum as a local professor, much more inexpensively. Colleges can change teachers as often as they want, creating a competitive market for adjuncts worldwide. Low student-to-faculty ratios correlate with better academic development,[47] and global adjunct professors can drop these ratios to unprecedented levels.

At Princeton the average student pays $2,000 a year for textbooks alone. Professors assign the books, but don't have to purchase them, so students get trapped in the microeconomic cage called "price inelastic demand." Students *need* these books in order to complete their course work, so publishers know they can sell the same number of texts for $200 apiece as they can for $100, and prices float upward. With flexbooks and public domain material, competitive universities can eviscerate these costs.

Classes can expand beyond campus. A group of students won't need a formal schoolroom in which to meet. They could have the same experience in a home, at a park, in a library, or hanging upside down on the jungle gym. Students show up with their tablets and leave with their tablets, and they won't need special power sources or wiring. The line between classroom and field trip can blur, and the cost of facilities can be trimmed.

Students can also gather in cyberspace. Online learning is an idea that has been failing, to a certain extent. But mobile technology enables onscreen video classrooms with teachers and students worldwide. As was said before, technology fails until it succeeds, and in the educational arena, it's poised for success.

What holds a civilization back? Ultimately, it's a hundred people doing the work of one. When we freed most farmers from the field, we delivered far more food to the people, at much less cost. When we freed most brokers from the office, we enabled far more investment—again at a much lower cost. And when we free most teachers from the classroom, we'll deliver far more learning, for much less.

So parents can *invest* their savings instead of writing checks for tuition, and students won't have to go into hock to gain a professional's synapses. The drop in college costs will usher more people into higher education and expand the worldwide knowledge economy. Ultimately, we'll look back on the current college system as we do the old brokerage business—wondering at its inefficiencies, and grateful to have escaped it.

Global Education

India's first Prime Minister, Jawaharlal Nehru, gave his nation a glittering gift. He founded five world-class universities—the Indian Institutes of Technology—and they prove the power of higher education. Many of their graduates immigrated to the United States, where they played high-profile roles in the fortunes of Silicon Valley and were indispensable in the rise of Indian high-tech firms like Infosys.

Yet India's adult literacy rate is 65 percent.[48]

Its schools excel at the top and struggle at the bottom. Why? The Brahmins in India's caste system has long pressured officials to focus on secondary schools rather than primary, since they themselves have private tutors for their young children.[49] At the same time, there are Indians battling to become primary school teachers. It's a nice job. The pay is good but the hours are better. A teacher doesn't have to show up for work at all, and some don't. The average public school teacher appears in class just 75 percent of the time, thus wasting 22.5 percent of the nation's education budget.[50]

Teachers are civil servants, and under Article 311 of India's constitution it's almost impossible to demote them, much less fire them.[51] So primary education languishes in India, though it always gives a higher rate of return to society than secondary education. For each dollar invested, it makes people more literate, more aware of health hazards, and more productive.

Elsewhere, corruption is blunter. The least literate region on earth is West Africa, where we see adult literacy rates of 23.6 percent for Burkina Faso, 28.7 percent for Niger, and 24.0 percent for Mali.[52] These remain oral worlds, largely cut off from the rest of the planet. And in 2010, Transparency International conducted a study of seven nations in Africa—Niger, Senegal, Uganda, Sierra Leone, Ghana, Morocco, and Madagascar, It revealed that 44 percent of parents had to pay bribes to get their children into school.[53]

The palm is out in universities, too. In Tajikistan, students may have to pay a $2,000 bribe in order to enter college, and some families sell their cattle to get the funds. In Turkmenistan, these bribes are called *elaklyk*, or "giving thanks."[54] Throughout Central Asia professors charge fees called *stavki*—ranging from $500 to $900—just to pass exams, so some students, even in medical school, don't bother with class. In Ukraine, 56 percent of students paid bribes to enter school.[55] One professor in the nation of Georgia actually handed out a list of his bribe prices to students.[56] Such corruption isn't just a Central Asian phenomenon. It occurs in Peru, Bangladesh, and everywhere there are less developed countries.

A buyable diploma is hardly a diploma at all, since bribes contaminate the very good they offer.

Mobile devices will broaden these markets, and shatter the incumbents. And students in Mali or Lima will gain not just a certified education, but a better, richer one. When a tin-roof hut can become a library carrel, and you can earn a more meaningful degree on a tablet than in a classroom, universities will be forced to crack down on bribes. They'll have no other way to compete.

Education commonly undermines corruption. For instance, the United States was awash in it during the nineteenth century. City dwellers relied on favors granted by ward heelers for jobs and business, and in return they backed operators like the fat, cheery William Magear ("Boss") Tweed. But as people grew better educated, they more easily found work on their own, and by the end of the century they were organizing against corruption, all in the backlash called the Progressive movement.

This is the sort of backlash that has occurred, in a more intense form, in Tunisia and Egypt—two of the best-educated countries in Africa. More such scenarios will continue to boil over across the planet.

They will occur because mobile technology is accessible to just about everyone—even the illiterate. The magazine that reads itself is not exclusive to commuters. With translation capabilities and text-to-speech, unschooled peasants on the Bolivian *altiplano* will have access to a *New Yorker* article, just as fast as Cambridge professors. They and their children will discover the modern information stream.

The next step is literacy.

Nearly a billion people entered the twenty-first century unable to read a simple paragraph,[57] and a tablet computer is better than a schoolhouse. The old ways of teaching are slow and expensive. They require teaching staff, paper, books, and a classroom where students meet during set times of the day. Adults may not be able to attend because of their work, and in many places children labor in the fields all day long. But with mobile technology, cost plummets, access broadens, and pedagogy rises.

Mobile devices have already spread throughout the developing world, and since users don't have to be literate to use them, all they have to do is touch the right icons, and study whenever they want. If children in northern Brazil have to work, and can't go to school, they can still learn to read and write in the evenings. Moreover, as tablets proliferate, females are more likely to become literate. Everywhere in the emerging world, they are less literate than men. But as prices fall and they acquire their own mobile devices, more of them will master the written word.

With video, lessons will spring to life. An instructor, standing there onscreen, can hold up steaming coffee, and then reveal the Swahili word *kawaha*, so the word imprints more deeply on the students' neurons. Immediately the viewers can practice writing it onscreen. With text-to-speech, they can listen to an article, and see which written words correspond to the spoken ones they know, and can stop at any time to investigate a word more thoroughly. They can learn through play—for instance, the Mobile and Immersive Learning for Literacy in Emerging Economies (MILLEE) program uses games adapted from popular culture, in order to heighten literacy among school-aged children.[58]

In 1956 the newly independent Sri Lanka replaced English as the official language, adopting Sinhalese instead. Nine years later, when Singapore began pulling itself up out of the Third World, it made English its primary language. Today, Sinhalese speakers number in the tens of thousands worldwide, and Sri Lanka's people live in a linguistic cul-de-sac. Singaporeans are hooked into the global language.

Literacy is a spectrum. The person who can read twenty-five words is still illiterate, just less illiterate than the one who can read none. And in our transnational world, ignorance of English is another degree of illiteracy. It limits a person's participation in the global conversation. But mobile excels at teaching English.

Immersion is the best way to learn a foreign language, and video can place anyone from an Egyptian farmer to a Siberian pharmacist into an English-language chat session. Digital text offers built-in dictionaries that not only define words and give examples, but show videos. Want to know what "liquidambar" means? A print dictionary will tell you it's a tree, a sweet gum, but a mobile dictionary will show the leafy ornamental plant quivering in the breeze.

Infants are already using tablets. A toddler in Sri Lanka can start learning English early, grow up one-hemisphere bilingual, and be connected to other youngsters across the world. The diversity of languages remains a costly barrier, and mobile will help break it down.

At higher levels, we can create an engineer for $200 a year, about the cost of a single textbook in the United States. And by sharing the tablet, students can cut this cost further. The poorest three billion people live on less than $900 annually,[59] so $200 is a price point that starts to make more sense—especially with well-targeted assistance. We can train other professionals the same way. For instance, by 2010, a Tanzania-based program called BridgeIT was already sending digital instruction to help train teachers in hundreds of schools in Tanzania and the Philippines.

"Give a man a fish and feed him for a day; teach him to fish and you feed him for a lifetime." With the exception of certain public health campaigns, foreign aid has failed for more than fifty years, as there are still 925 million hungry people worldwide.[60] Donations of food today raise the question of where tomorrow's food will come from, and consumable aid of any kind is probably the least leveraged way to help the impoverished. But give them a tool by which they can learn, build a trade, and make money, and you set economies in motion.

It's like space flight, which all comes down to the propulsion force. We won't see millions of people lift off into outer space unless we invent an engine a thousand times more efficient than a rocket. But in the information economy, we already have one—the CPU. It's raising the intelligence level of everyone on the planet, and is about to propel forgotten provinces and villages into the global conversation. With the computer, we enable the great adventure of mass education in the Third World.

CHAPTER 9

DEVELOPING WORLD

Bootstrapping the Developing World

Zinder, Niger – Mobile Networks: Instant Internet Infrastructure –
Mobile Technology Will Increase Market Efficiency – Eliminates
Middleman Costs – Makes Financial Services Available, and –
Overcomes Corruption – The New Globalization – Mobile Technology
Unleashes the Latecomers Advantage

A thousand years ago, Zinder was an outpost on the lonely route across the Sahara Desert. Now it's a city of 200,000 people in southern Niger. It has seen the rise and fall of sultanates, a clutch of revolts, and the French conquest of 1899.

But its most radical event occurred in 2003.

It takes a special tourist to visit Niger, one of the poorest nations on earth, and Zinder doesn't even lie on Niger's meager tourist trail. Despite its population, visitors find just a scattering of small hotels and few landline phones. Camels and donkey carts plod down the streets. The city has no supermarkets or ATMs. The electricity comes and goes. The water teems with parasites, and Zinder province has a 62.7 percent rate of the waterborne eye disease, trachoma—the highest rate known to epidemiologists.[1]

The market is the economic heart of the city. Every Thursday, merchants gather to trade the essentials of life. Desert Tuaregs trade Sahara

FIGURE 9.1
Cell phone availability in the third world makes the markets much more efficent by enabling farmers, like the man above, to know the daily market prices before taking his live stock on the long trek to market and by cutting out the middle men.

salt as they gobble roasted locusts, and Hausa tribesmen drink camel's milk yogurt as they bargain. For grain farmers, the trip to market takes two to four hours, and it's a big gamble, since if they don't find the buyers or the prices they need, the trip is pointless.

In Zinder—and throughout rural Africa—communication has been spotty. More than 55 percent of sub-Saharans listen to radio broadcasts at least once a week, using it as their primary source of information. Fewer than 20 percent read a newspaper, since papers are costly and hard to find, and much of the population can't read at all. Landline phones and fax machines are as rare as Hummers.[2] So if a farmer has wanted to know who would be at the Zinder Thursday market, he's had to go there himself.

In 2003, the company Celtel built a mobile-phone tower in Zinder. Soon, bright kiosks dotted the city, selling prepaid phone cards.

"I can get information quickly and without moving," a wholesaler said at the Thursday market. And a grain trader noted that, because of his cell, "I can know the prices for just $2, rather than traveling (to the market), which costs $20."[3] With his mobile phone, he could talk to buyers and sellers, find the best price, and place orders.

By providing basic person-to-person communication, the simple cell phone streamlined the Zinder market. However, this improvement pales compared to the changes that the cell tower will bring to Zinder—and hundreds of cities like it in the developing world. The cell tower will gradually open up the Internet to everyday people in Zinder as they acquire smartphones and tablet computers. Internet access is a game-changer for these communities—it will have huge positive impact that will ripple outward.

Mobile Networks: Instant Internet Infrastructure

When a single 3G or 4G transmitter station with a forty-mile range is built in a city, that city is catapulted straight into the twenty-first century.

Road networks, railways, banking networks, water systems, electricity, school systems, and health care networks—no other core infrastructure has ever been so rapidly deployable. They have been economically impractical to recreate in developing countries. And creating communication networks used to be equally impractical. The cost of laying copper wire in a city of 200,000 like Zinder would be more than $50M for the wiring alone, and at least another $50M in labor and equipment charges to bury it in trenches. (Stringing wire on poles would be infeasible due to the theft of the copper.)

The cell tower infrastructure can be deployed for tens of thousands of dollars, making it a thousand times less expensive.[4]

Modern 3G/4G wireless networks are revolutionary because, in addition to providing voice communication, they link people to the Internet at very high data rates. However, without cost-effective computers to leverage these networks, the full potential would go unrealized. Personal computers with 3G cards were impractical for many reasons. They cost up to twenty times more than cell phones. PCs draw fifty to one hundred watts of power, compared to cell phones at 3.7 watts (while charging). In fact, if only 10 percent of the population in Zinder had PCs, those units would require 1,500KW of power. That's more than is used in *all* of the Zinder households today.[5]

The same number of mobile devices would consume only 74KW, or 5 percent as much as the PCs. Moreover, PCs require a steady supply of electrical power to be useful. Mobile devices with their batteries can easily accommodate the frequent residential power outages experienced in developing areas. Because mobile computing allows independence from electrical infrastructures, third world countries can connect to the global economy with a dramatically lower cost of entry. For a few hundred dollars each, residents can acquire the "infrastructure" needed to come into the First World.

So, the poor worldwide have snapped up mobile technology. A number of developing countries—such as El Salvador, Venezuela, Guatemala, Panama, South Africa, Gabon, Armenia, Malaysia, and Thailand—have adopted mobile communications faster than the United States.[6] There are 4.3 billion people on earth who have adequate sanitation,[7] but 4.6 billion own mobile phones, and two-thirds of them live in emerging markets.[8]

India and China added 300 million new mobile subscribers in 2010,[9] a number nearly equivalent to the entire population of the United States. At last count, China had 420 million on the Internet, and 277 million using the mobile Internet.[10] By the time you read this book, the number of mobile phone users in China will exceed the number of U.S. citizens.

Mobile Technology Will Increase Market Efficiency

In the dark hours of the morning, in the Indian state of Kerala, sardine fishermen board their crafts and set forth onto the choppy Indian Ocean. The larger boats can each hold three dozen crewmen, and when they spot a glimmering school of sardines, they toss a five-ton net overboard, then haul it in.

Back on shore, a day's catch can be worth $220 or more, a significant sum in this area.

Prior to 2001, these fishermen faced a vexing problem. As captains returned to shore, they had to choose among the beach markets to sell their catch. These markets were about ten miles apart. They could visit only one market, because the fish were perishable and the markets closed at 8 a.m., but the price at each market could vary widely. Indeed, one market might have no buyers at all, and the fishermen would have to dump their whole catch into the sea, while just a few miles away, captains sold every fish, and some buyers left empty-handed.

Economists call such cases "Coordination Problems" where buyers and sellers cannot match up efficiently.

Mobile phones reached Kerala in 1997, and gradually penetrated society. Since most of the state's people live near the ocean, cell towers went up along the shore. By 2001, 60 percent of fishermen had cell phones and could use them as far as eighteen to twenty miles out to sea. An information network arose on the water. Now fishermen always know the best markets to head for, and indeed they could make their deals even before they reached land.

"When I have a big catch, the phone rings sixty or seventy times before I get to port," a captain of a seventy-four-foot vessel said in 2006. Price differences have almost vanished, and they no longer dump their fish. Their profits have increased by 8 percent, and consumers pay 4 percent less for sardines.[11]

Mobile phones "coordinated" this market. They fused the different beach markets into one and provided price and demand transparency to all of the participants. Buyers and sellers competed across wider territory, and prices evened out. Bigger markets are more efficient, and since both buyers and sellers benefit, they generate greater aggregate wealth for the community. More than 70 percent of Kerala's population—or 23.4 million—eat fish every day. The 4 percent consumer savings on fish provides the population with the equivalent 341 million days of free fish each year.[12]

Mobile technology is generating these market efficiencies throughout the Third World. Take Kenya. In the past, that nation has been a mosaic of isolated markets where merchants and farmers couldn't see beyond their own local horizon. One market might have high prices, and another low.

But in 2011, farmer William Muriuki showed a reporter a glimpse of the new world. He entered "price cabbage embu" using SMS on his cell phone. A reply came back swiftly: "Cabbage Ext Bag 126kg selling at Ksh400 in Embu as of 2011-04-01." That is, a bag of cabbages weighing 126 kilograms had sold for 400 Kenyan shillings (US $5) in his local Embu market during the past week.

Then he typed "price cabbage Nairobi." A reply came: "Cabbage Ext Bag 126kg selling at Ksh2100 in Nairobi as of 2011-04-01." Ksh2100 was equal to $26.26, or over five times higher than the price in Embu.[13] So, transportation costs permitting, he would choose to sell in Nairobi. As more and more Nairobi farmers track their local markets this way, prices for goods will continue to fall. Farmers and city dwellers will grow richer.

As app-phones and tablets replace cell phones, the markets will become further streamlined. Captains won't have to handle sixty to seventy calls each morning; the prices will be available at a glance. The market will expand beyond the beach markets to include anyone in cyberspace who wants to bid.

In Kerala, mobile technology went further than just improving market efficiency. It improved productivity while cutting risk. Captains learned good areas to fish each day. They could return to shore early, if supply grew high and the price was low. They got weather forecasts—especially wind speed and wave height—and knew whether they and their crafts would be safe. These forecasts generally reassured fishermen, brought more boats onto the ocean, and boosted the catch.

In emergencies such as engine breakdowns, fishermen could contact shore and save lives.[14]

On land, mobile devices can also make every acre more bountiful. They give Third World farmers access to online markets for seeds, herbicides, and pesticides. They provide weather forecasts, so growers avoid fertilizing or planting just before a storm that would wash everything away.[15] They provide longer-term forecasts so farmers can decide which crops to plant. In 2010, Africa had approximately 137,000 cell towers[16] and each one was an ideal site for an automatic weather station.

And when goats would sicken, tomatoes become discolored, or coffee berries shrivel, farmers could snap photos of them and send them to experts to get advice.[17]

Mobile Technology Will Eliminate Middleman Costs

As a consequence of market inefficiencies, the Third World suffers a major drain in its economic value chain: the middleman.

Middlemen coordinate markets. They bring sellers together with buyers, and take a cut for themselves. Without them many markets wouldn't function at all. But whenever they can be eliminated, both the producers and the consumers save. In opaque markets, middlemen can know the crucial details and exploit both buyer and seller. These opportunists abound in Third World agriculture, and mobile technology is poised to sabotage them.

Consider India.

In 2009, 52 percent of its labor force worked in agriculture,[18] but as of 2010 they generated just 18.5 percent of its GDP.[19] The nation was plowing a disproportionate amount of human intelligence and energy directly into the dirt. Farmers—especially the more isolated ones—have been astoundingly ignorant about every stage in the market process, from foreseeing consumer preferences to obtaining seeds to selling their produce effectively. They haven't known the resources available outside their local areas, and they've lacked social networks to discuss issues with other farmers. They haven't known the current prices at the end of the value chain, and they may not understand the chain at all.[20]

They've been perfect marks for middlemen.

And India has more or less institutionalized the middlemen. For decades, most states have had laws that compelled farmers to sell to a government agency, except for small amounts they were permitted to sell directly to consumers. The statute arose to protect farmers by providing transparent prices, but that goal got lost in the bureaucratic maze that spawns corruption. So an array of middlemen game the system.

And it's big business. One study compared two traditional vegetable markets in Ahmedabad and Chennai with a small *non*-traditional market in Chennai where farmers sold directly to citizens. In the traditional markets of Ahmedabad and Chennai, farmers received between 40 percent and 69 percent share of the consumer price. But in the non-traditional, direct-to-consumer market they got 85 percent to 95 percent.[21]

It happens in every major town in the country. The farmers get less than fair market value, while the citizens pay higher prices. Without middlemen acting as intermediaries, farmers would have more money to invest in machinery and fertilizer, and consumers would have more money to spend on other goods. India's GDP would rise, and more people could get out of agriculture and provide higher-order economic value to society.

Many Indian states have now altered their monopolistic agricultural statutes in order to let farmers sell directly to city retailers and enter into private contracts. Forward contracts lock in prices for the farmers, safeguarding them from the vagaries of weather and other variables of farming. Moreover, clauses in corporate contracts often stipulate that the company will provide technological aid to the farmer—an arrangement that benefits both parties. The National Spot Exchange formed in India in 2008, where farmers could sell their produce without middlemen.

Opening up the flow of information was a key factor in all of these changes. Mobile technology provided the information flow in the rural areas that were least able to afford a more traditional communication infrastructure.

Mobile Technology Makes Financial Services Available

On September 15, 2008, Lehman Brothers filed for bankruptcy and triggered a credit crunch that sent stock markets plunging and destroyed companies worldwide. As the credit multiplier effect reversed, the globe entered the worst recession since the 1930s, and the slow climb out of the recession became partly a matter of restoring trust, and of freeing up credit for people to use.

Credit lubricates the entire economy. It not only lets people survive income gaps, but also allows them to buy and grow. Without credit, you can't gauge the cash flow you'll have in the future, and turn it into current opportunities for investment. For instance, if no one could get a mortgage, no one could buy a house unless they had saved up the money for it. So the housing market would compress. If no one could get a loan for college, higher education would compress. In both cases, people would lose jobs, less money would circulate, and we'd sink in toward deflation.

Cut off credit and you cripple the economy.

If you make a precipitation map of the earth, the low numbers describe the polar areas and the great wastelands above and below the tropics—the Sahara, the Kalahari, the empty quarters of Australia. The zones are fairly regular.

If you create a credit map of the earth, the low numbers form a far less regular outline: their boundaries outline the Third World.

The Third World is a credit desert. Worldwide, 53 percent of adults or 2.46 million are unbanked. They lie outside the formal financial system. The highest rate is 80 percent of adults—326 million people—in sub-Saharan Africa. But there are also lofty percentages in the Middle East (67 percent), Latin America (65 percent), East Asia (59 percent), and Southeast Asia (58 percent). In high-income countries, by contrast, only 8 percent of adults are unbanked.[22] In India, 18.3 million people possessed a credit card as of March 2010.[23] That's 1.5 percent of the population: virtually no one. Most people there use cash, with all its friction.

When they need loans, they rely on informal sources like friends, money lenders, shopkeepers, and landlords. In one study of thirteen developing nations, the poor borrowed just 6 percent of their funds from formal sources like banks—except in Indonesia, which had recently launched a large microcredit program. Most of the rest of the money came from informal sources, and those loans can be costly.[24] In Hyderabad, India, for example, the average informal interest rate was 57 percent per year. And that's typical.[25]

It would be as if most people in the United States got their loans through check-cashing services.

If you want to start a business in the Third World, you use your own capital and loans from friends. In one study of 14,000 micro businesses in Mexico, 61 percent of the founders used their own savings, and 14 percent used those of friends or family.[26] Even after a business is established, formal loans may remain out of reach, though they get easier over time.

Does extra capital make a difference to these small firms? In one study in Sri Lanka, experimenters chose a random sample from 408 small businesses and gave them $100 each. The companies that got the money increased their monthly profits by $38.50 to $53, which corresponded to an increase in the annual return to capital, from 55 percent to 63 percent. A similar study in Mexico found even higher returns, with increases ranging from 900 percent to 3,000 percent per year.[27]

Since banks in Sri Lanka charge 12 percent to 20 percent per year, you'd think that they would be eager to lend to small firms. Indeed, if loans were available at 20 percent, entrepreneurs would be lining up and banging on the door, yet just 3 percent of these firms even had a business bank account.[28]

The fundamental problem is transaction costs.

The process for conducting transactions costs too much. The bank needs information—such as the borrower's address, occupation, income, references, and record of paying loans in the past. If the bank charges 20 percent for a one-year loan of $100, its transaction costs can't exceed $20, or it can lose money just making the loan. If banks try to raise the interest rate, the default rate rises, too, so banks shrug when entrepreneurs come calling.

As a result, businessmen turn to more costly ways to borrow. Hence, credit moves sluggishly or not at all across whole economies. The Third World financial pump doesn't reach most of the body.

Microfinance banks have solved the problem—but only partly. Microfinance has been around at least since satirist Jonathan Swift, who created the "Irish Loan Fund System" for poor farmers in the eighteenth century. In Bangladesh, Muhammad Yunus opened the Grameen Bank in 1983, which has since loaned some $8 billion, and for his efforts he won the Nobel Peace Prize in 2006.[29] His banks offer low interest rates and survive because they cut transaction costs and default rates. For instance, they seek less background information. They also dangle the reward of a larger loan if borrowers repay the existing one promptly.

And they tend to loan to groups, so there is social pressure to honor the debt. Grameen Bank has a 98.6 percent repayment rate and claims that 64 percent of its borrowers with accounts five years old or older have moved above the poverty line.[30] But, its profits have always been slender, and Yunus has viewed the endeavor as philanthropic.

Credit will penetrate the Third World when it becomes clearly commercial.

A mobile phone is a channel to drive credit. For instance, tablets make it easy to open bank accounts, even small ones. These accounts can include all your finances and other records, so banks have transparency and can issue loans at reduced cost. In addition, as banks automate and simplify processes, they can hire less skilled workers to collect credit, whom they can pay less. So the cost of issuing a loan will drop further.[31]

If Third World banks prove slow off the marks, the capital-rich First World may have a business opportunity coming down the mobile channel. Legal restraints aside, there's no reason why a merchant in Zinder couldn't have an account with Bank of America. Such arrangements would benefit the Third World far more than most foreign aid. They would be ongoing and systemic, and would encourage entrepreneurial energy.

Mobile devices would also spur use of digital cash. The greatest success so far has been the Kenya program M-Pesa (in Swahili, "M" for "mobile" and "Pesa" for "money"). M-Pesa is a system for depositing, withdrawing, and transferring money. Registration requires just a national ID card or passport. Perhaps wary of regulation by the Central Bank of Kenya, M-Pesa doesn't pay interest or make loans. Nevertheless, it has proved extremely popular. More than three quarters of all Kenyans have access to mobile phones, and in a land of 10 million households, there were 14 million M-Pesa accounts by 2011. They held 40 percent of the country's savings, and some people used them for all their shopping.[32]

"M-Pesa" had become a verb, as in "I'll M-Pesa it," and the average transaction was tiny—about one-seventieth the size of the average check in Kenya.[33] Mobile payment systems have arisen in South Africa, the Philippines, and elsewhere in the Third Word, but M-Pesa is the most developed.

Mobile can make similar benefits available with insurance, again by cutting transaction costs. A farmer might pay $1 to get $10 of insurance against drought. But if a drought occurs, a claims adjuster has to visit the farm to check the damage, and the company has to fill out paperwork before issuing the payment. So a $1 premium won't even cover the administrative cost. And since that cost is fixed—it's about the same for a $1 premium as a $1,000 premium—companies have focused their services elsewhere.

But mobile can short-circuit the process. In southwest Kenya, the program Kilimo Salama, or "safe farming" in Swahili, was insuring 22,000 farmers by 2011.[34] The company sold policies in feed stores, where the owner snapped a photo to record the purchase and a message went by text to the customer's phone. Instead of checking individual farms for drought, the company hooked each client up to a nearby, computerized weather station. If the station registered low rainfall, all farmers linked to it got automatic payments, which varied by rainfall amount. No one had to file a claim, and the sum went straight to the farmer's M-Pesa account.[35]

Transaction costs shrank. The most notable cost of this program was the welcome email the company sent to the customer.

As more people become creditworthy, the cost of goods will fall. The more insurable they become, the more sound businesses will be saved. People will take more reasonable risks, just like the fishermen with their weather forecasts. Entire societies will benefit.

Mobile Technology Overcomes Corruption

In 2002, a reporter for *The Economist* began a four-day journey through Cameroon in a sixty-ton truck with 30,000 bottles of Guinness beer. By the time the trip ended, police had halted it forty-seven times and stolen a third of its cargo.

The roadblocks were usually tires or oil drums, and the officers commonly engaged in leisurely, microscopic inspections of tires, taillights, and wing mirrors. At one stop, the reporter heard that his visa was on the wrong page of his passport, and at another that he lacked the right *number* of permits—at which time the officer offered to sell him one.

One policeman told the driver he had broken the law against carrying a passenger and impounded his driver's license. Informed that no such law existed, he patted his holster.

"Do you have a gun?" he said. "No. I have a gun, so I know the rules."[36]

As blatant and crude as such corruption may be, it's nothing compared to the spectacle of a Mobutu Sese Seko of the former Zaire, who extracted $5 billion out of his impoverished, mineral-rich nation—or Marcos in the Philippines, or Suharto in Indonesia, or Zimbabwe's Mugabe. It is, however, more dramatic than the everyday corruption that becomes a part of the environment—people's way of doing business because everyone *else* does business that way.

Such routine corruption is everywhere in the Third World. For instance, in 2005, the World Bank found that it took two days to approve a business in Toronto, but 153 days in Maputo, Mozambique. Registering a commercial property took three procedures in Helsinki, but thirteen in Abuja, Nigeria.[37] Complain about red tape in these places, and the palm comes out. Throughout the developing countries, people give bribes to have garbage men collect trash, druggists fill prescriptions, bosses release salary checks, and police decide against false arrest. Paying for such services is paying for nothing.

This is anti-infrastructure.

As funds earmarked for infrastructure vanish en route, the price to improve education, transportation, and communication keeps increasing, and economies remain listless. As public services become less available, taxes can rise, so more people evade them, state revenues fall, and shadow economies—black markets—appear. Companies waste valuable management time dealing with the prehensile officials, and they pass the costs on to the consumers. And though Third World countries badly need foreign investment, multinational companies enter warily, knowing that officials will gnaw at their returns and make profits less predictable.

In other words, corruption correlates inversely with growth.

The African Union estimates that corruption costs their continent some 25 percent of its annual GDP, and the figure is around 15 percent in Mexico.[38] The World Bank notes that people pay around $1 trillion in bribes every year,[39] and most of these deals are in the Third World. In contrast, a key secret of Singapore's success lay in killing corruption from the start, and straight dealing has also helped the economies of Chile and Botswana.

Since the turn of the century economists, non-governmental organizations, and politicians have focused increased attention on corruption, yet it has not declined. It's a complex problem, especially when it's pervasive. For instance, it's hard to make an example of a cheater if everyone knows a hundred other cheaters who are getting away with it.

But bribery lives in the shadows, and mobile technology is illumination. Official corruption becomes easier to document and reveal, and governments come under greater pressure to intervene. Social networks also let people share concerns about, for instance, the cash some Indian nurses demand from new mothers before they can see their babies. In groups, people can support one another, encourage whistle-blowing, and take more effective action.

Mobile technology also lets governments sidestep their own freebooting functionaries, by providing online services with standard procedures that cut out parasitic middlemen. For instance, to register land holdings in the Indian state of Karnataka, citizens once had to deal with village accountants—a process fraught with delays that only bribes would end. In 1998, the state launched a program called Bhoomi, designed to computerize land records. Within a decade it had registered 20 million holdings. Since the process is automatic and employed statewide, it has bypassed the accountants and reduced bribery (though some has remained, because it didn't bypass revenue inspectors).[40]

Regardless, studies indicate that automating government processes cuts corruption in general.[41]

Mobile technology introduces other shortcuts that serve to bypass corruption. Mobile banking services like Kenya's M-Pesa can automatically deposit salaries in employee accounts, so they don't have to pay the bosses just to receive their checks. Mobile technology enables more people to track expenditures for projects like roads and hospitals, so leakage diminishes. Mobile technology also puts isolated people into contact with others in countries where bribery is rare. Once they understand that it doesn't *have* to be a part of the fabric of life, they are encouraged to change the norms.

Yet, corruption affects even mobile phones. In 2008, governments in sub-Saharan Africa were collecting tax on mobile operators, equivalent to 30 percent of their revenue.[42] Governments can also thwart competition by refusing to grant radio spectrum licenses to new participants. But the problem of corruption is ultimately one of misaligned incentives, and they tend to adjust properly as nations grow more affluent.

The New Globalization

From 1750 to around 1950, the economies in First World nations grew at about 2–3 percent a year.[43] Higher rates seemed impossible on a steady basis. But since then, nations such as China have racked up rates around 9 percent, year after year.[44] This growth astonished economists at first, and yet it could go higher.

The road from Third World to First is paved with exports, and countries such as Japan developed coordinated catch-up strategies focused on global trade. First it developed petrochemicals and steel, then shifted to downstream industries like shipbuilding and autos, and finally moved into plastics and electronics. In consumer goods it followed a stepwise pattern, starting with simple goods like apparel and moving to more complex ones. The idea was for Japanese companies to sell abroad at a profit, and bring capital into the country—which they have. So have companies in Taiwan and China.

Mobile technology will accelerate this process. It will make it much easier for buyers and sellers to find the best prices worldwide, much as the fishermen did in Kerala.

If you were a Third World farmer in the 1980s, and wanted to launch an export business shipping corn to the First World, you'd need information on crop protection, taxes, tariffs, accounting, currency hedging, and shipping. And you'd require a sophisticated infrastructure, such as banking and insurance—not to mention the phone lines so often stolen for their copper.

Today, in theory, you could get communication lines via your tablet without anybody having to string copper wires to your house. You could do the banking off the tablet. You could do the shipping off the tablet. You could get the crop protection off the tablet.

Mobile technology enables clearinghouses—initiatives such as Amazon.com's Mechanical Turk, which taps into cheap labor overseas for micro-tasks like transcribing audio or identifying objects in a photo. It can pay a person in a Dhaka slum a few dollars a day—a big improvement if you've been living in poverty.

Global trade will extend deeper down into the small business realm than ever before. On a worldwide basis, micro and small businesses far outnumber large companies and account for a major chunk of GDP. As mobile devices penetrate these layers, such operations will expand the international networks of information, production, and transportation. Globalization itself will evolve.

And mobile technology will breed more micro-multinationals—startups with no nationality at all. These firms can be very small and extremely nimble, with perhaps five or six people spread across three or four nations. When companies cover the world, they can tap into the varied market strengths everywhere. For instance, the image-tagging company Viewdle has two people in Uruguay, four in California, and three in Ukraine. Founder Laurent Gil, who is French, noted that his company invented the technology in Ukraine, got capital in California, and found good engineers in Uruguay.[45]

Moreover, a virtual firm like this can save enormously on rent, since it doesn't need office space. In the dotcom era, some startups in San Francisco's Mission District were paying $77 a square foot and filling their offices with $1,300 Herman Miller Aeron chairs, pool tables, and "romp spaces." The Canadian telecom firm Nortel turned a suburban factory into an office structured like a mini-town, complete with streets, parks, cafes, and a Zen garden.[46] This was lavish waste—often of someone else's venture capital. Dotcom entrepreneurs sought to remake the office. Today's tech startups seek to eliminate it.

In addition to the cost of rent, offices abound with productivity drags. These include interruptions, small talk, political infighting, and often-needless meetings. Employees waste time and energy in commuting, and they pay for fuel and wear-and-tear on their vehicles. Older people consider offices more important, probably because they're more used to them, but globally many new entrepreneurs lack that mentality. The varied costs of an office can be the difference between profit and loss, success and failure—especially at the start.

When the first corporations arose in the nineteenth century, they needed offices, because people had to be close to one another in order to communicate. Today they can carry the office in their pockets.

Mobile Technology Unleashes the Latecomers Advantage

Together with exports, a rising nation needs technology. Countries such as South Korea were relatively backward in the 1950s, as veterans of the Korean War will recall. But such nations didn't have to invent, refine, and develop technology such as computers, because others had already done the work. They simply had to acquire it—which they did, sometimes by illegal means like copying, and sometimes by licensing and foreign investment.

This quick, cheap infusion of technology plainly contributed to South Korea's 9 percent growth rate.

Moreover, these countries didn't bear the burden of antiquated technology, such as primitive adding machines. Economists call this lack of antiquated infrastructure the "latecomer advantage." Mittal Steel illustrates the effect. It began as a small company in Indonesia in 1976, and didn't start expanding internationally until 1989. With that late start, it could use mini-mills and electric arc technology from the outset, without having to bear the cost of replacing obsolete operations. One by one, Mittal bought state-owned enterprises across the Third World, and when it had constructed a global network, it purchased companies in Europe and the United States. In 2004 it became the world's biggest steel company.

Today, mobile technology is introducing shortcuts we've never seen before. Market entrants have the greatest latecomer advantage in history.

In 1900, much of rural America was technologically crude. It had no electricity, phone service, or running water, and almost no cars sputtered along rutted roads.

In the ensuing years we paved roads and built the interstate highway system. We developed cheap air travel and efficient container ships. Suburbs, strip malls, and branch banks spread. So did book store chains, libraries, large hospitals, and schools. Radio, TV, and personal computers appeared and became ubiquitous, as did credit card readers and ATMs. We laid millions of miles of copper, coaxial, and optical fiber from coast to coast. White-collar work became synonymous with fax machines, FedEx, phones, copiers, paper, printed reports, broadband, and the Internet.

If a price was placed on all of this infrastructure, it would represent the barrier to entry that traditionally stood between a town in rural Africa and entry into the First World. However, an emerging nation doesn't need to acquire all of those things. In essence, mobile phones are a massive progressive subsidy for the poor, a gift of more than 130 years of investment. A sleepy place like Zinder cuts the structure out of infrastructure and enters the twenty-first century directly.

The transition from Third World to First is a complex process with multiple, interacting strands. It can involve the internal markets, middlemen, and corruption, as well as demography, religion, local strife, national industrial policy, and many other factors. Mobile technology, alone, won't necessarily transform a nation. But, as First World services come free of their physical anchorages and start to flow as digital streams, it will be realistic for those services to find their home anywhere a wireless signal can reach.

Even the *Wall Street Journal* can now be delivered to a ghetto in Mumbai. Or a trader in the grasslands outside Zinder could place profitable bets on credit default swaps or foreign exchange, employing the same information as a person based in lower Manhattan.

CHAPTER 10

NEW WORLD

Human Energy Unleashed

The Agricultural Revolution: From Fields to the Cities – The Industrial Revolution: From Farms to Factories – Reassigning Human Energy – The Information Revolution: Information Everywhere – The Economies of Revolutions – Change Will Come Fast – Privacy in the New World – New World, New Rules – Final Word

Mobile technology will change how people conduct their day-to-day lives.

It will change how businesses operate, and it will change entire industries and the economies they power. With so much change, is it appropriate to call this a "Mobile Revolution?" Is it akin to the Agricultural Revolution that transformed humanity from nomadic groups of hunter–gatherers into city dwellers, creating the foundation for the great city–states of Greece and Rome? Is it like the Industrial Revolution that ushered in the modern mechanized economies we see today?

I believe that mobile computing is the tipping point technology for the larger Information Revolution. That revolution started with the printing press in the 1400s, but it wasn't until the advent of computing technologies in the 1960s that the Information Revolution began to accelerate its impact on society. Mobile computing is the fifth wave of computing technology, and it will be the catalyst that brings society the most dramatic changes of the Information Revolution.

The Agricultural Revolution—From Fields to the Cities

The agricultural revolution "began" as early as 10,000 B.C. when hunter-gatherer nomads first began to settle into semipermanent villages. The "technology" that triggered humans to put down roots was the domestication of plants and animals, which obviated the need for everyone to spend their time seeking it out. However, it took another 6,000 years before the fruits of the Agricultural Revolution revealed itself in the form of the first true cities in the fertile crescent of Iraq.

In fact, Iraq probably got its name from the first great city on earth, Uruk, which was situated on the Euphrates River within the borders of modern-day Iraq. By 3,200 B.C., Uruk was the pinnacle of human economic complexity. Boats glided down canals past large temples, citizens appreciated the metalwork along lanes containing a myriad of mud-brick shops, and artisans created statues in their workshops. On the quays, stevedores unloaded ore and timber brought from the Iranian mountains. Beyond the city walls, laborers tended irrigated fields, date groves, and orchards. In the evening, citizens sipped beer and were entertained by lyre music.[1]

At this time, Uruk supported 50,000 people and incubated human innovations like the first writing, schools, and large-scale civic institutions. City residents traveled shorter distances for their amenities, enjoyed better information, and in their markets they had better choices, higher quality, with lower prices.

Socio-economic revolutions like the Agricultural Revolution revolve around harnessing new sources of energy. The incredible diversity of economic actors in Uruk—farmers, tanners, builders, weavers, metal craftsmen, brickmakers, boat builders, scholars, politicians, shopkeepers, and candle makers—was possible only because new agricultural technology created a surplus of bio-energy—food. The surplus freed up human energy—people—that could be turned toward higher-order economic areas such as crafts and shop keeping. The greater

FIGURE 10.1 Water mills, an immovable source of energy, powered the days before the Industrial Revolution.

man. Water power and wind power machines existed with some regularity throughout ancient Greece, Rome, and China.

Watermills and windmills powered grain mills for grinding flour, they powered sawmills for cutting trees into planks for shipbuilding, they powered textile mills with trip hammers for pounding fabric, and they powered steel mills by pumping the air bellows that fired the furnaces.

It's not surprising that these early industries all had the term "mill" in their names. One doesn't hear about "car mills" or "shoe mills." New energy sources would be needed to propel these new industries.

Making Mechanical Energy Transportable: The Industrial Revolution was sparked in the 1700s by the emergence of the first reliable

complexity of the economy increased the circulation of mon
everyone became richer.

As less human energy was needed to provide bio-energy f
population, no doubt someone asked, "What will all these peop
who no longer need to farm?" The answer is obvious in hinds
They became the foundation of the middle class of craftsman (i
manufacturing sector) and merchants, shopkeepers, and bureau
(in the services sector).

Having surplus human energy is a key to human advancement
many places of rural Africa, women walk miles every day to fe
water for their families. The average household spends 134 minu
per day on this chore, with related costs in time, calories, and diseas
Yet in cities with indoor plumbing technology, the trek disappear
time frees up, and people are more prosperous.

Interestingly enough, humanity did not advance markedly from
those early urbanization years. The standard of living across the cities
of the medieval Europe maintained arguably the same or lower standards when compared to the cities of Mesopotamia, Egypt, Greece,
Rome, Incas, Aztecs, and China. Even though worldwide populations
increased markedly over the thousands of years from 3000 B.C. to
1700 A.D., the per capita gross domestic product (GDP) remained
frozen at about $600, as expressed in 1985 dollars.[3] Each additional
human just added his increment of $600 to the worldwide GDP.

It wasn't until the industrial revolution in the 1700s that human
productivity and income received a major boost forward with the beginning of the Industrial Revolution.

The Industrial Revolution—From Farms to Factories

The Industrial Revolution was also about harnessing energy—in this
case, mechanical energy. Prior to the beginning of the Industrial Revolution in the mid 1700s, water power, wind power, and muscle power
(from beasts of burden) were the primary motive forces available to

steam engines. The steam engine had the almost magical property of being able to convert one form of energy—heat provided by the burning of coal or wood—into another form of energy—mechanical movement. Previously, if someone wanted mechanical energy, they harnessed it from something that was already moving—like water, wind, or a horse.

Despite the introduction of steam power in the 1700s, water, wind, and muscle continued to be the primary sources of mechanical energy for manufacturing and farming. Steam didn't replace wind and water power. Instead steam introduced the radical new idea that mechanical energy could be "transportable."

With steam, a manufacturing operation didn't have to be situated permanently next to a river or on top a windy hill to get power. Instead, it could generate mechanical power anywhere. All it needed was a boiler to heat a tank of water, a piston assembly that was pushed by the pressurized steam, and enough levers and gears to make the moving piston do something mechanically useful.

Steamships plied the waterways carrying their power with them; steam locomotives transported previously unimaginable weights of cargo across continents; steam shovels dug canals that connected the Gulf of Mexico with the Pacific Ocean; and steam pumps emptied water from coal mines allowing more production of this high-energy material that would be used for powering more steam engines.

Watermills were to the Industrial Revolution what mainframe computers are to the Information Revolution. Both were powerful and essential, but stationary. By contrast, steam engines were the minicomputers of Industrial Revolution. For the first time, people had transportable mechanical power, in the same way that minicomputers gave us transportable computing power.

Making Mechanical Energy Portable: Oil was the next major advancement that would propel the Industrial Revolution forward, enabling the creation of gasoline and diesel engines. These new power plants were like the personal computers of the Information Revolution. They took the portability of power to a whole new level, leading to new applications like automobiles, trucks, motorcycles, and chainsaws. These were applications on a more personal scale. Industrialists in the early 1900s probably scoffed at the idea of gasoline engines. They were far too small for anything useful in a factory where large-scale water and steam power ruled. However, new smaller technology often doesn't replace bigger predecessor technologies. Instead, they bring new and unexpected applications that are different than anything envisioned in the previous worldview.

Mainframe and minicomputer manufacturers scoffed at the first personal computers, as well. They considered PCs to be toys.

Placing Mechanical Energy Everywhere: Electrical energy is the tipping point technology of the Industrial Revolution. It put machines in every corner of modern society, filling almost every niche of mechanical motion needs, ranging from large-scale stationary operations in industry and homes powered through electrical networks, to transportable operations powered by generators, to portable operations like car batteries, and ultimately to mobile operations with lithium ion batteries. Everywhere you look, you see machines driven by electricity.

Electricity has become the universal power platform of the Industrial Revolution. It allowed mechanical power to become ubiquitous. We still use oil, natural gas, water, wind, and nuclear fuel, but most often we use those fuels behind the scenes... to produce electricity. In the same way, mainframe computers, minicomputers, and personal computers will continue to be used in the Information Revolution, but they will be used primarily behind the scenes as "servers" that provide information power to mobile computers. Mobile computing will become the universal computing platform of the Information Revolution, and will be the tipping point technology that allows software to become ubiquitous.

Reassigning Human Energy

In 1840, 69 percent of the U.S. labor force worked in farming, creating food for a population of 17 million.[4] These farmers had little time to do anything else—like write a book or run a factory.

But breakthroughs from the Industrial Revolution—including tractors, harvesters, fertilizers, and new livestock breeding techniques—changed everything. By the year 2000, just 1.9 percent of the U.S. labor force worked in farming, yet we *still* were creating a surplus of food.[5] By freeing up time for 98 percent of workers from farming, we grew wealthier and we liberated human energy that could be used for building cars, constructing cities, and developing computers and the software to run on them.

As of 2010, 36.7 percent of the worldwide labor force worked in agriculture.[6] That equates to 1.18 billion people out of a worldwide labor force of 3.23 billion. At the same time, in the United States, agriculture labor became even more efficient, with just 0.7 percent of the workforce providing our food.[7] If the same level of food productivity could be achieved on a worldwide basis, then only 23 million people would be required to feed the rest of the planet, freeing up 1.16 billion people to do something of higher economic, intellectual, or artistic value than cultivate food.

The 2010 data shown in Figures 10.2 and 10.3 show how far out of balance the worldwide use of human energy is in comparison with its economic contribution. The U.S. numbers are surprisingly balanced, with approximately 1 percent of the agriculture labor force contributing approximately 1 percent of the country's GDP. Approximately 20 percent of the manufacturing labor force contributes 22 percent of the country's GDP. About 79 percent of the services sector labor force contributes 77 percent of the GDP. This is almost a perfect balance of labor energy and production value.

The worldwide numbers are not nearly so balanced. A disproportionately large percentage of people work in agriculture (37 percent)

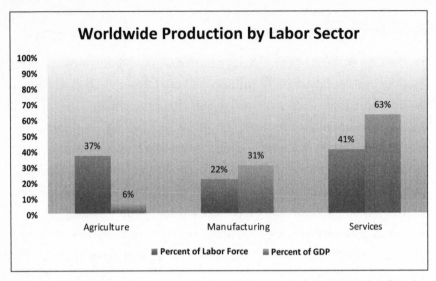

FIGURE 10.2 The agriculture sector represents 37 percent of the global labor force but only produces 6 percent of the world's GDP. This is a great imbalance.

Data Source: Data from 2010. See "World FactBook."

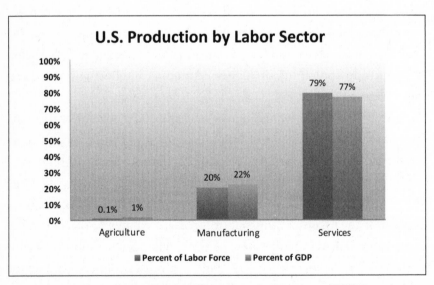

FIGURE 10.3 The ratio of percentage of laborers to the percentage of GDP is much closer to 1 for the U.S. This is a better balance than the world as a whole.

Data Source: Data from 2010. See "World FactBook."

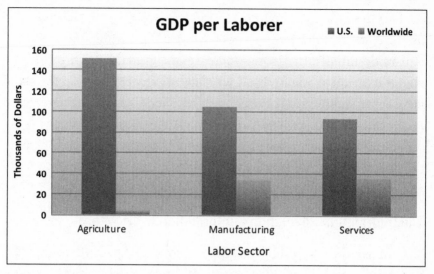

FIGURE 10.4 In all three primary labor sectors, especially in agriculture, the U.S. has much higher labor productivity (GDP/Laborer) than the world as a whole.

Data Source: Data from 2010. See "World FactBook."

and that massive labor force contributes a very small amount of GDP (6 percent).

Figure 10.4 shows the massive difference in productivity for the United States, versus the entire worldwide average. Each U.S. worker produces between 93–150 thousand dollars of GDP, with farming leading the GDP contribution with 150 thousand dollars per person. By contrast, worldwide farm workers only contribute 3.8 thousand dollars of GDP. That differs from the U.S. by a whopping factor of 40. Worldwide workers in the manufacturing and services sectors only contribute between 33 and 38 thousand dollars of GDP per year respectively.

The agricultural and industrial revolutions clearly shifted the allocation of the U.S. workforce from agriculture to a near 80/20 mix of services and manufacturing. It has shifted the rest of the world's workforce, too, but the process is clearly not complete—there's a lot more friction to overcome.

This next wave of the Information Revolution promises to shake up the distribution of labor at every layer. While mobile computing will undoubtedly fuel some increases in global agricultural and manufacturing productivity, it will have a greater effect on the services sector where it will eviscerate inefficiency and automate away many low-skilled services jobs. Just as previous revolutions reassigned human energy for positive benefit, so too will the Information Revolution.

The Information Revolution—Information Everywhere

The Information Revolution might be considered to have begun with first cuneiform writing in the fourth millennium B.C., but its first mass impact on populations didn't occur until the appearance of Gutenberg's printing press in the 1400s. The revolution extended its reach and immediacy with the advent of the telegraph and telephone in the 1800s, and it was amplified via the mass communication of radio and television in the mid-1900s.

Nevertheless, it wasn't until the advent of commercial computing in the 1960s that the Information Revolution found its steam engine.

Each new computing wave brought new applications that increased productivity. The "mainframe wave" automated bookkeeping and tabulating functions within businesses and government. The "minicomputer wave" automated factory processes. The "personal computer wave" automated office work with email, spreadsheets, and word processing. The "Internet wave" automated many consumer-to-business interactions with online ordering and customer service.

Just as the harnessing of electricity was the tipping point for the Industrial Revolution, with mobile computing, we now have the tipping point technology for the Information Revolution. Mobility makes all previous applications more valuable by virtue of its perpetual nature (24x7), accessibility (wireless networks), usability (apps), and availability (lower cost).

A New Force: Information Energy: Like the other revolutions before it, the Information Revolution has energy at its core. The Information Revolution is about harnessing "information energy."

Information energy is not a common phrase, but most people can grasp it intuitively. It is the fuel that drives people—and machines—to make a decision and take a course of action. By analogy, if coal is the fuel used by steam engines to create the mechanical motion that drives a steamboat into action, then information is the fuel used by computers to create "decision motion" (apps and reports) that drives people to a course of action.

Information energy is not like other physical energies so one needs to be careful in drawing incorrect analogies. Unlike physical energy, there is no concept of conservation of information energy. In fact information can be consumed, but it cannot be depleted. It can be used over and over again, infinitely, to productive purposes. Information can be created but never destroyed, unless it's lost or forgotten. In fact, every action (and inaction) creates new information, so it is constantly growing.

Unlike oil, every different "drop" of information has a different energy value. The "house on fire" drop has more energy than the "air temperature at 68 degrees" drop. One of these information droplets drives much more action than the other. Even the *same* drop of information can have an entirely different energy value depending on who is consuming it. For example, today's dollars-to-euro exchange rate has far more energy value to a currency trader than to a taxi driver.

Another difference is that oil energy is purely additive, such that ten drops of oil contains 10 times more energy than one drop of oil. The energy content of information, by contrast, is exponential. This means that the energy content of a collection of data drops may be much higher than the sum of the amount of energy contained in each individual drop. *Together*, they raise the stakes. For instance, total wheat production for a single year is valuable information; but total wheat production for ten years, combined ten years of rainfall data, and ten years of fertilizer represents thirty times more data droplets, but probably contains one hundred times more information energy, because it shows trends and correlations that will drive a greater number of decisions.

Information Energy Removes Friction: We use information all the time to conduct our professional lives. Bankers use fiscal data to decide whether or not to grant a loan. Insurance underwriters use risk information to assess policies. Doctors use EKG information to decide on a course of heart treatment. Caregivers use school dismissal alerts to know when to pick up their charges at the bus stop. Drivers use traffic information to pick the fastest route. The volume of decision making and coordination that takes place at every level of our modern society is simply staggering. Ever since our move from farms to cities—with their complex webs of interactions—information has been the key to coordinating every aspect of society.

Information doesn't just affect our professional lives. It also permeates our daily personal lives. Consider how much time is spent by people coordinating carpools, making reservations, identifying relevant news, planning social events, shopping, and juggling deliveries. All of these efforts and many more—often including the mundane—are targets for mobile applications today. The Information Revolution is all about reducing time, streamlining processes, cutting out the middlemen, enhancing coordination, and increasing efficiency.

Information Energy—Vast Untapped Potential: Information is being collected at a staggering rate by the computers, software, and sensors that run our modern societies. Cash registers record every transaction down to the specific types of food one buys. Internet commerce systems maintain a complete history of every purchase you make and can track the products you've looked at. Sensors on roadways monitor traffic. Bar-code scanners record the movement of every tube of toothpaste from manufacturer to your shopping bag. Stock markets track every trade and price with perfect fidelity. Aerial reconnaissance measures crop health. Real estate systems record the details of every transaction. Power meters record how much energy our homes use. Even our cars are recording the health and status of their engines.

The more computers we put in place to help run things, the more they will record what is happening.

Businesses are creating data at an astonishing rate as a byproduct of their information systems. In 2011, Wal-Mart was handling over a million transactions with customers every hour.[8] Large firms like eBay, Bank of America, and Dell each managed data volumes in the petabytes (one million billion bytes, or 10^{15}), and global data has reached the zettabyte level (one trillion billion bytes, or 10^{21}).[9] Social networking is contributing a vast amount of additional data, as well. In 2011, Facebook had more than 50 billion photos stored in its records, seven for every person on earth, and each month people shared 30 billion items.[10]

Twitter was relaying 155 million tweets a day.[11]

Every minute, YouTube stored twenty-four hours of new video.[12]

The total global volume of data has been growing at 40 percent a year.[13] We are witnessing, not a flood of data, but a supernova. Mobile computing has the potential to take this even higher.

Before the dotcom revolution, many of the most ambitious databases in business only tracked purchase transactions. With the Internet, web retailers could not only track every item people bought, but also could track the entire sales process. Internet retailers could answer questions like: "How many TVs did customers look at before they chose the brand they bought?" They could identify televisions that are so strongly branded that people purchased them with zero comparison browsing. Suddenly, automated retailers had critical information about the basic nature of products, and about consumers' decision processes.

Now multiply that level of automation by a factor of ten, or a hundred, as mobile computing automates more and more aspects of our personal lives. We will use apps to purchase products, pay bills, make reservations, hail taxicabs, apply for visas, and pay train fares. We will use mobile phones to snap more pictures, post them, and interact with friends while we are on the go. The GPS chips in our phones are useful for displaying maps, but they are equally good at reporting our position. It could accumulate an immense amount of data about crowd concentrations and movements. It could pay for itself by telling retailers when to expect high foot traffic, and supply personal data about those who are passing by.

In the mobile world, systems will be capturing far more information about us on an individual and group level.

The shocking thing about information, however, is not how much there is, but how inaccessible it is despite the immense value it represents. Various technologies are advancing to tap into this value. Database technologies are advancing to handle the larger and larger volumes of it. Business intelligence technologies are advancing to extract more data and present it in more consumable ways. Applications and websites are being created that streamline and automate processes.

What has been missing is a delivery device that is ubiquitous. The web browser interface on desktop computers emerged in the 1990s as the most widespread and commonly available interface, so it evolved as the primary interface for large-scale applications. However, desktop computing allows information usage only during times when a person is sitting at their computer. Yet peoples' decision-making and coordination hours span their entire waking day.

Mobile computing puts information energy in hands of individuals during all waking hours and everywhere they are.

The Economics of Revolutions

Where and how will the mobile-driven phase of the Information Revolution affect economic structures?

Nowhere are the benefits of harnessed information energy more important than in the services sector of the economy. This includes banking, insurance, education, health care, retail, restaurants, real estate, media, entertainment, lodging, transportation, law, government, and telecom, to name a few. Services industries do not produce tangible products, like manufacturing industries, or produce food, like the agriculture industries. Instead, they exist to facilitate commerce and human interactions.

Automating the Service Sector: To a very large extent, service industries are people-powered. People execute the services whether driving a truck, making a medical diagnosis, writing a contract, approving a license, operating a cash register, or managing a hotel desk. Industrial machines whose primary function is to move "things" have been ideal for automating the many moving parts of manufacturing and agriculture. But such movements cannot automate loan approval, or teaching geometry, because there is nothing tangible to move.

Service activities like loan approval and teaching primarily involve moving information. So *information* machines are needed to automate the service sector.

The service sector is the largest economic sector in the U.S. by an overwhelming margin. As of 2010, it employed 79 percent of the work force or 122M people, and it accounts for 77 percent of the GDP or 11.28T dollars.[14] This very high contribution is consistent among other high-income economies where food and product needs are largely already met, and people increasingly look to intangible services to improve their standards of living. The question this begs is, what will become of so many workers whose jobs are automated away by new mobile apps? What happens to the librarians, the clerks, the waiters, the loan examiners, the lawyers and doctors, the real estate agents, insurance agents, and bureaucrats whose primary job function is to serve up information, or who simply move information from one outbox to another inbox?

The answer is the same as it has been through all previous economic revolutions. This surplus human energy will ultimately be reapplied to higher-level economic activities, many of which haven't yet

been discovered. When farmers moved to the city during the Industrial Revolution, no one could have conceived of the diversity of jobs
that would be waiting for them. The same was true for factory workers moving to office jobs at the beginning of the Information Revolution. A librarian today whose job is eliminated by mobile books might
be employed tomorrow by Google tagging the world.

Through each wave of farm automation, more and more farm
hands moved to manufacturing. Through each wave of manufacturing
automation, more and more factory workers turned to the services
sector. Figure 10.5 shows the effects automation has had on this labor
migration from the goods producing sector to the services sector. In
1970, the ratio of service sector workers to goods producing workers
was 2.2 to one. By 2000 the service-to-goods worker ratio rose to 4.4
to one, and in 2005 it was five to one.[15]

With information automation, service workers will migrate their
human energy to new and more specialized service areas. More and
higher-level services will require a greater level of education, and this
is likely where much of the higher per capita income will go. Economists theorize that populations will continue the trend of shifting
family resources from producing *more* offspring to making certain
they have offspring who are *better prepared* to react to the new
economies.[16]

Fortunately, education should become much more available, and
much more affordable.

To Create, You Must Destroy: Looking at just the United States economy, one might theorize that 10 percent of the services sector could
be streamlined in the next five years due to mobile applications and
computing. That implies that 10 percent of the current human energy
applied to the services sector might be freed up, or approximately 12
million people. While this isn't a staggering number, it's a politi-
cally poisonous number given the large-scale unrest witnessed in 2010
with its 9.6 percent unemployment rate equating to approximately 15

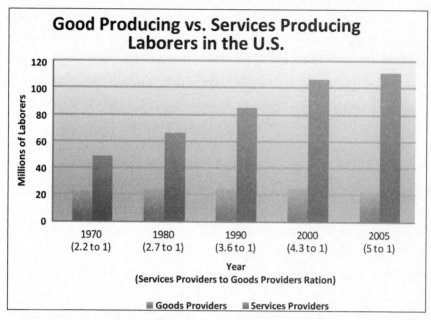

FIGURE 10.5 Over the past thirty-five years, the number of workers in the services sector has increased while the goods producing sector has decreased. Increased manufacturing automation forced factory workers to move to the services sector. Information automation will create a new shift among service workers, a shift to more specialized service areas.

Data Source: See "The American Workplace - The Shift to a Service Economy."

million workers. But this calculation ignores the new businesses and new activities that will be created.

The introduction of the Internet itself can serve as an example of what we can expect with mobile computing.

In May 2011, McKinsey Global Institute published a full economic report on the impacts of the Internet.[17] According to the report, the Internet had an undoubtedly positive impact on global growth. In developed countries, it accounted for 21 percent of GDP growth over the last five years, and traditional industries received 75 percent of the benefits. But once again, what happened to those workers who lost their jobs due to new Internet efficiencies? The McKinsey report claims that the Internet created 2.6 jobs for every job it eliminated!

In another report, McKinsey analysts determined that the explosion of data, which they call "big data," would create a need for 1.5 million managers in the United States.[18] That's not all. The Internet has boosted the productivity of small- and medium-sized businesses by 10 percent, and the heavy users have exported and expanded twice as much as others. The more a nation uses the Internet, the wealthier it becomes.[19] If our experience with the Internet is any gauge of the economic future of workers displaced by services automation, then mobile computing will be far more of a benefit, than a problem.

Automation has created increasing wealth for the world as a whole. Worldwide wealth, measured as GDP per capita, experienced 0 percent growth during the time period from 1 A.D to 1700. Worldwide production grew at almost the exact same rate as worldwide population, stagnating worldwide wealth.

During the dawn of the Industrial Revolution in the 1700s, worldwide production finally grew faster than population for the first time in the last 2,000 years, with a meager 0.3 percent growth in GDP per capita. Nevertheless, people were finally getting wealthier. During the age of steam power in the 1800s, production grew discernibly faster than population with GDP per capita growth averaging 1 percent per year. During the petroleum era of 1900–1960, GDP per capita grew at 2.4 percent per year. And in the beginning of the information age, from 1960–1990, GDP per capita grew an average of 4 percent per year.

"For the first time in history, the living standards of the masses of ordinary people have begun to undergo sustained growth," Nobel winner Robert Lucas said in referring to the growth experienced throughout the industrial revolution and especially at the beginning of the Information Revolution. "The advances in technology increase the wages of those who had the skills to put these technologies to economic use. These wage effects stimulated others to acquire these skills." Essentially the shoemakers and the farm hands accumulated new skills and became members of the vanguard for the new technology.[20]

Change Will Come Fast

Each revolution comes faster than the previous one. The Agricultural Revolution took *thousands of years* to transform vast tracts of land, populated by nomads, into cities, towns, and villages around the world. The Industrial Revolution imposed its massive socio-economic changes over a period of a few *hundred years*. By simple extrapolation, we might expect that the Information Revolution will require *tens of years* to achieve pervasive changes in our lives and businesses.

Any prediction, however, is almost guaranteed to be proven wrong over time. Nonetheless, it's reasonable to expect that by 2018, we will see a massive upsurge in innovation relating to mobile applications throughout the developed world, and great penetration in the developing world. By 2025, we will see almost universal use of mobile computers as our primary means of navigating through modern society.

I'm not the only one who is making these predictions. Cisco Systems, the technology company that helped define networking in the Internet age, sponsored a comprehensive global study to examine the growth in mobile data traffic over the past few years, and with these findings it modeled predictions for usage in 2015. The facts are unquestionably stunning and a clear indicator of the onset of the mobile revolution.

Over the past three years leading up to 2010, mobile data usage has nearly tripled from one year to the next, and by 2015 mobile data traffic will increase twenty-six-fold from its usage in 2010. Over the past year, smartphone users have on average doubled their individual data traffic, and by 2015 this will increase sixteen-fold.[21] So as a smartphone user, in the year 2015 you will use sixteen times the amount of data you used in 2010. These numbers are clear indications of how mobile technology has become an integrated part of daily life, and shortly will be becoming an essential part of life.

Advances in wireless data technologies have contributed greatly to the above increases. Consider that 4G LTE (Long Term Evolution) service offerings that hit the market during the 2010 time frame offered peak download rates of 100Mbps, a five-to-ten-times improvement over prior 3G technologies. A 4G upgrade, LTE Advanced, is anticipated for 2012 and promises peak download rates of 1Gbps, yet another order of magnitude improvement. These dramatic speed improvements, coupled with the usability and affordability of smartphones and tablets, will drive more than 788 million people to use mobile devices as their only source of Internet.

Currently there are 48 million people in the world with cellular phones, but no electricity,[22] and by 2015 we will see a world with one mobile device per capita.

Privacy in the New World

Henri Cartier-Bresson gained international acclaim for his photos of lively Parisians who seemed oblivious to his camera. In fact, they *were* oblivious, since he roamed the city with a handkerchief hiding the device. Was he acting unethically? We sense an invasion, but his motive wasn't to harm anyone. And if you snap a stranger's photo and show it to a friend, any damage is usually trifling. But what if Cartier-Bresson caught a compromising moment, as through an open window? He could own the image permanently, so there *would* be some harm, and if he sold it to a tabloid there would be far more.

In fact, the camera spurred the first strong arguments for a legal right of privacy. In an influential article, Samuel Warren and Louis Brandeis urged the protection in 1890, noting that photography and the rise of big newspapers posed a novel threat, and stating that courts had to evaluate privacy in light of "modern enterprise and inventions."[23]

By 1960 we had developed essentially the privacy rights we have today.

Modern enterprise and inventions have by now far outstripped the newspapers. Even infants are photographers, webcams and networks are ubiquitous, and people upload millions of new photos every day. When I can take hundreds of shots of you and display them to every human on earth in perpetuity, I can find a way to make you look ridiculous or disgraceful. If you happen to burp, say, millions in Europe and China can see it, again and again. If just 1 percent of the 155 million daily tweets are defamatory, that's 1.5 million instantly re-tweetable instances of libel, so the damage can swiftly exceed calculation. You may not know whom to sue, but even if you do, court costs are high, the process can take years, and no award may suffice in the end.

And if you were in a "public place," you would have no legal protection at all, except in the case of commercial uses of your image.

Viral communication also changes the meaning of "publication," and therefore privacy. In September 2010, a female student at Duke created a forty-two-slide presentation as a mock senior thesis describing her sex life—"horizontal academics," she called it—with thirteen college athletes. She sent it to three people, who found it amusing and forwarded it to others, including alumni overseas. As it went viral, two websites published it, and it spread to millions. She had expected the privacy of a small group of friends, but the audience grew out of control.

Nobody was prosecuted, no one went to jail over this incident, yet the young woman's reputation was wrecked, as were the lives of these athletes. It wouldn't have happened twenty years ago, because no newspaper of any repute would have run the story. But today, our seemingly private exchanges are all one click away from becoming global theater.

Do we have a right to privacy, even when in public? Suppose you're walking into a fertility clinic, and I stand outside, snap your photo, upload it, and tag you. Is your right to a private medical visit greater than my right to broadcast it? What if I snapped photos outside an AIDS testing center, or a drug treatment clinic? You're in open view. And the photos have the potential to be circulated far and wide.

In theory, someone could take Google Street View to the next level: Street Watch. Cameras could be placed on every street corner, stream data up to the Internet, and run photo imaging and facial recognition software. It could continually tag people—without their knowledge or consent. It would be the ultimate stalker device. Anyone with access could type in your name, and see everywhere you've been, outside of your home or office, in the past twenty-four hours. Eyes in Sri Lanka or Kazakhstan could follow you. And not just eyes. Someone might set up a sensor that listened outside your house.

Most people consider such activities as far too invasive, and thus most people acknowledge a right to at least some privacy, even in public. Temporary exposure to a few other pedestrians is one thing; permanent exposure to billions is another. Privacy laws still ignore the difference, however. That's why paparazzi can snap photos of you for tabloids.

And governments could commit even worse abuses. The leaders of a nation might say, "We're going to track your location every hour of the day, for the common good, and we'll use this information to make the trains run on time, do urban planning, and improve national defense." And perhaps use it to prevent the spark of a revolution, like the uprising in Tahrir Square in 2011 in Cairo.

Yet governments themselves are vulnerable, too.

In November 2010, Wikileaks released approximately 250,000 U.S. State Department cables, in an action that violated the privacy of a wide variety of policymakers. For instance, the public learned that the State Department worked with the brother of Afghanistan president Hamid Karzai, though it deemed him a corrupt drug trafficker. We learned that the Saudis wanted the United States to bomb Iran. We learned that U.S. diplomats called French President Nicolas Sarkozy an "emperor with no clothes," and Russia's Vladimir Putin an "alpha-dog."[24] Some revelations, such as the latter, were simply remarks in private that became indiscreet when laid bare.

We need a new movement, like the one spurred by Warren and Brandeis. We've forgotten why our privacy laws arose, and we need to re-examine the basic principles of privacy and craft a new theory for privacy and dignity for the individual. Traditionally, governments are slow to grasp these matters, and indeed Washington seemed to view the Wikileaks affair as a national security concern, rather than a more fundamental one.

The answers are very complex and demand not only a deep consideration of conflicting goals, but also a base-level rethinking of digital issues themselves. For instance, does it now make sense to consider privacy a form of property? Perhaps each individual has a slender but real trademark-like interest in his or her own image. If so, what "fair use" rights do others have to reproduce it without consent, as for instance when a foul ball comes down next to a spectator at a baseball game and TV cameras capture that person's face?

Should we treat the great software networks as regulated utilities? Software economics tend to concentrate them into a few winners, and we now depend on them. They give us our electricity and our health care. As they provide basic services like the code to open your door, or to buy groceries, we'll depend on them much more. So should the government regulate them?

And what are the First Amendment hazards of regulating an information utility, as opposed to one that just provides gas?

Many people have suggested an electronic bill of rights, and the time is overdue for an intelligent, thoughtful legal model of these rights. The issues were inconceivable in Constitution Hall during the summer of 1789, but they are all around us today. Even a respected but non-binding legal model, such as the Restatement of Torts, might be of tremendous value in our modern society.

New World, New Rules

Each revolution creates an upheaval requiring new rules and new cultural dynamics. Information automation will lead to the loss of anonymity. There will always be a trail of who you are, where you go, and what you do. Even as we establish rules to protect that data, and limit its use, it will still exist. So the bottom line is that it will be a matter of information forensics to dig it up.

There will be other issues, beyond the question of privacy. There will be services or products available only to those people with sufficient computing power, and unavailable to those who do not have computing power. It might be that some hotels will only serve people with proper "key software" because it enables them to operate without clerks, and perhaps enables them to offer a safer hotel. As retailers stop accepting cash, only those people with the right software will be able to shop in a store that has adopted such a policy.

We will need to be wary of companies that gain monopolistic power over life's essentials. Software has an odd property among products in that people really don't want choice and diversity. They want one or two products in every category, so they can focus their attention. Imagine what it will mean, however, when one or two companies control a core process in daily life. For instance, in 2000 you didn't rely on a software company in order to have friends. In 2011, you do— for millions of users, if Facebook cut you off, your life would be thrown into turmoil. Facebook has become the standard for social sharing.

Consider public safety. Suppose someone creates "safety net software" that monitors police scanners, reports your location on a regular basis, collates safety status reports for locations all around a city, and dispatches help the instant you press a distress button on your phone. Sounds incredibly useful. You're safer. And as more and more people have this sort of software, it becomes essential. But what happens when the organization that operates that safety network controls it for an entire region. Everyone depends on that organization for safety—without them, there is none.

In banking there's a concept of "float" for funds that are briefly neither in the payer's nor the payee's account, because they are in transition. If a person has control of the float of, say, $10 million and creates a three-day delay, that person might use that money and enjoy a lucrative payoff. So, will there be a float on information? If a person knows your location thirty-two seconds or thirty-two hours before anyone else, the float might be valuable.

As more and more information flows throughout society, we may have to consider physical concepts like easements, entitlements, and trespass, and apply them to information rights. And novel issues will arise.

Gutenberg's printing press brought problems, too. Scholars predicted it would spread nonsense—like astrology—far and wide. And it did. It bred libel and insurrection, as well. But it also spurred science, quickened markets, and laid the foundation for our modern world. Mobile technology raises new and difficult challenges, but it, too, offers vast benefits, and we can enjoy them fully if we also mitigate the drawbacks.

Final Word

The projections described in this book are logical extensions of today's technologies, propelled by basic forces of economics. If a new thing is technically feasible and is far more economical than the old thing, then the new thing will happen—sooner or later.

What makes the mobile revolution so compelling is that it is actually more than just technology and economics. It's positive psychology—mobile apps are fun.

It's pervasive—billions participate in it.

It's empowering—software extends the individual's capacity and influence.

It's ethical—mobile levels the playing field.

The next generation will grow up immersed in mobile computing. They will see it as normal and will wonder why people carried around wads of animal hide to hold paper money, credit and loyalty cards, and identification. Take a poll of every parent you meet and ask when they intend to get their child a mobile phone. The answer used to be age sixteen, when they started to drive. Then the answer was thirteen, when they went to middle school and needed to be picked up from after-school activities. Now it's primary school. Soon, parents will need to get phones for their first-graders, so they can purchase food at the school cafeteria. Children might begin to receive their phone numbers at birth, to act as a universal identifier, and they will likely keep those numbers for their whole lives and use them to draw social security someday.

The next generation will wonder about quaint old customs like reading the morning paper, standing in a line, checking in and out of a hotel, going to a store to shop, sitting in a library to study, using the office copier, and waiting for someone to un-jam the copier. The corner newsstand and the airport bookstore will be symbols of a bygone era, like the phone booth today. The ATM machines will disappear along with bank branches and wallets. Our most ordinary and repetitive routines are likely to be the first to be automated away, because that's where we waste our time, and that's where the efficiencies will be.

Personally, I look forward to the mobile information revolution. It will be one of the most compelling periods in history. It will be one where each human becomes empowered to shape his or her environment and to enjoy access to the world's wealth of information and opportunities. I will miss thumbing through my collection of books, and I will miss seeing my neatly arranged array of CD/DVD jewel cases. But I will enjoy the convenience of having so many physical objects become software on my mobile phone. These virtual keys, wallets, and maps will perform functions their physical counterparts could not.

Software-based objects won't rust, they cannot be misplaced, and they can't wear out. Most intriguing to me, as a software entrepreneur and executive, is that this new world can be built with a minimum of capital, so it will be a market available to the every software designer in the world. We will see an explosion of creativity like never before in history.

Technology is acid. Unleash it and it burns away accumulated inefficiencies in economies, in industries, and in products. It dissolves extraneous links in production chains, shortening them, eliminating costs and time. It cuts away layers of middlemen and exposes corruption so that everyone can see it. The acid of technology etches away the unnecessary, reshaping things and leaving behind only the core that is durable enough to withstand it. Mobile technology is very powerful acid. It has been unleashed and it will change everything.

NOTES

CHAPTER 1

1. See "Global Mobile Statistics 2011: All Quality Mobile Marketing Research, Mobile Web Stats, Subscribers, Ad Revenue, Usage, Trends... | Mobi-Thinking."

2. See "Global Survey Shows Cell Phone Is 'Remote Control' for Life: 42 percent of Americans 'Can't Live Without It' and Almost Half Sleep with It Nearby."

3. See Hasay.

4. See "Statistics | Facebook."

5. See Shah.

6. The actual quote is slightly less memorable. Trotsky said, "Burnham doesn't recognize dialectics, but dialectics does not permit him to escape from its net." (See Trotsky.)

7. Google was worth $190 billion (see "Google Net Worth 2011."), and Facebook had an IPO valued at over $100 billion for the first quarter of 2012 (see Golgowski).

CHAPTER 2

1. See Morris.

2. See "Sabre: The First Online Reservation System."

3. See Cortada.

4. See Pelkey.

5. See Polsson.

6. See Licklider.

7. See Thomas.

8. See Schlesinger.

9. See Ahonen.

10. See "BlackBerry Users Hit by Eight-hour Outage — CNN."

11. See Lashinsky.

12. See Isaacson, p. 466.

13. See Guglielmo.

14. See Epstein.

15. See Gonsalves.

16. See Fox.

17. See Chmielewski.

18. See Van Der Velden.

19. See Plowright.

20. See Hotz.

21. See "IPhone4S Special Event 10/4/11 — Scott Forstall Demos Siri Voice Recognition — YouTube."

CHAPTER 3

1. See "AbitibiBowater Files for Bankruptcy Protection."

2. See Abramovitz.

3. According to research by the trade group National Association of Printing Leadership. See "Twenty Rugged Survivors in Dying Industries: The Commercial Printer: Suttle-Straus — BusinessWeek."

4. The percentage of consumers buying Christmas cards fell from 77% (see Jones) in 2005 to 63% (projected) in 2011 (see Hill).

5. See "Employment by Industry, Occupation, and Percent Distribution, 2008 and Projected 2018."

6. See "OECD Environmental Outlook (2001)." p. 218.

7. See Green Press Initiative.

8. See Martin.

9. See Oregon Department of Environmental Quality.

10. See Kramer, p. xxi.

11. See Pettegree, p. 4.

12. See Fuchs.

13. See Pettegree, p. 4.

14. See Manguel, p. 135.

15. See Temple, p. 81.

16. See Battles, p. 39.

17. See Temple, p. 83.

18. See Pacey.

19. See Martin.

20. See Yu, p. 7.

21. See Goodrich.

22. See Wong.

23. See Füssel.

24. See Eisenstein, p. 46.

25. See Robinson.

26. See "Gutenberg's Legacy."

27. See Brown, p. 184.

28. See Febvre.

29. Based on an estimate of 89 million people in Europe by 1600. See North, p. 103.

30. See Bawden, p. 75.

31. See Yu, p. 7.

32. See Dittmar, p. 1.

33. See "Frequent Questions | Paper Recycling | US EPA."

34. See above.

35. See Manguel.

36. See Finkelstein, p. 33.

37. See Finkelstein, p. 35.

38. See Meissner.

39. See Rich.

40. See "US Stats Show 9% Ebook Share, Grim News for Print."

41. See Battles, p. 30.

42. See McCollan.

43. See Sarno.

44. See National Center for Education Statistics, cited in Sarno.

45. See "Project Gutenburg."

46. See Crawford.

47. See "Case: NFC in Oulu City Theatre."

48. See "Choose Cremation."

49. See Blodget.

50. See Kirchhoff, p. i.

51. See Flores.

52. See "WAN — Newspapers: A Brief History."

53. See "WAN — A Newspaper Timeline."

54. See Kirchhoff, p. 3.

55. See Hurter.

56. See "The New York Times Company Reports 2010 Third-Quarter Results."

57. See Kirchhoff, p. 1.

58. See Kirchhoff, p. 4–5.

59. Newspapers' revenue from classified ads fell from $19.6 billion in 2000 to $6.2 billion in 2009. See "Annual (All Categories)."

60. See Edmonds.

61. See Edmonds.

62. See Zacks Equity Research.

63. See "Decline in Newsroom Jobs Slows."

64. See Cody, p. 14.

65. See "Frequent Questions | Paper Recycling | US EPA."

66. See "Hippopotamuses: Fast Facts."

67. See Halter.

68. See Kiehl.

69. See "Frequently Asked Questions.

68. See Kiehl.

69. See "Frequently Asked Questions."

70. See Martin.

71. See Martin.

72. See "Frequently Asked Questions."

CHAPTER 4

1. See "Flickr.com Site Info."

2. See Selburn.

3. See Pew Research Center, p. 17.

4. See Zhang.

5. See Purewal.

6. See Nickinson.

7. See above.

8. See Wortham.

9. See Valjalo.

10. See Arlotta.

11. See "Timeline of Computer History."

12. See Schonfeld.

13. See "Nielsen Says Video Game Penetration in U.S. TV Households Grew 18% During the Past Two Years."

14. See "Console Wars | Seventh Generation | Worldwide Sales Figures."

15. See "Mobile Apps Convert a New Generation of Mass Market Casual Gamers — Gadgets & Tech, Life & Style — The Independent."

16. See "Nielsen Wireless Survey, March 2010." p. 21.

17. See Ingram.

18. See "Most Innovative Companies: Zynga."

19. See Upton.

20. See Mangalindan.

21. See Meyers.

22. See Clifford.

23. See Krupa.

24. See Idvik.

25. See Anton.

26. See Chmielewski.

27. See Global Mobile Statistics 2011.

28. See O'Brien.

29. See above.

30. See Freeman.

31. See Clark.

32. See Schneider.

33. See Wesch, p. 84.

34. See "Deloitte's "State of the Media Democracy" Survey: TV Industry Embraces the Internet and Prospers."

35. See Netburn.

36. See Rose, p. 230.

37. See Burgan.

CHAPTER 5

1. See "Business Card History."

2. See Smith.

3. See Landrum.

4. See Valdes-Dapena.

5. See Valsic.

6. See "Zipcar to Replace City Vehicles in New Deal Saving $400K."

7. See Yule.

8. See U.S. Bureau of Engraving and Printing.

9. See Wang.

10. See "Fun Facts About Money."

11. See above.

12. See Karmin, p. 70.

13. See Voorhees.

14. See Sims.

15. See Martin.

16. See "The United States Mint · About Us."

17. See "United States Circulating Coinage Intrinsic Value Table."

18. See Ahern.

19. See Preson.

20. See "Making a Dollar Bill Now Costs 9.6 Cents."

21. See Weatherford, p. 210.

22. See Bellens.

23. See Jimenez, p. 2.

24. See "Global Mobile Statistics 2011: All Quality Mobile Marketing Research, Mobile Web Stats, Subscribers, Ad Revenue, Usage, Trends… | Mobi-Thinking."

25. See Heatwole.

26. See Shin.

27. See Woolsey.

28. See Evans, p. 4.

29. See "Moving Assembly Line at Ford."

30. See Calder, p. 40.

31. See Evans, p. 54.

32. See "2008 Worldwide Credit Cards Usage Statistics and 2009 Trends."

33. See Evans, p. 4.

34. See Petruno.

35. See Frellick.

36. See Calder, p. 11.

37. See "Apple Event."

38. See Tuttle.

39. See Dukes.

40. See "The Lowdown on Customer Loyalty Programs — Forbes.com."

41. See Keiningham, p. 116.

42. See Vavra.

43. See Keiningham, p. 112.

44. See Cox.

45. See Gilfoyle.

46. See Truman, p. 1.

47. See "Official Identity Theft Statistics | SPENDonLIFE."

48. See Foley.

49. See Zetter.

50. See Heaton, P.5.

51. See Truman, p. 1.

52. See Truman, p. 2.

53. See Clark.

54. See "In the Face of Danger: Facial Recognition and the Limits of Privacy Law."

CHAPTER 6

1. See El-Heni.

2. See Hölldobler.

3. See Vermeij, p. 21.

4. See Nakashima.

5. See Chafkin.

6. See Adegoke.

7. See Calvo-Armengol.

8. See Schonfeld.

9. See Carter, p. 32.

10. See "Cell Phones Key to Teens' Social Lives, 47% Can Text with Eyes Closed."

11. See Lenhart.

12. See Ghelawat.

13. See "Eminem | Facebook," "Rihanna | Facebook," "Lady Gaga | Facebook," "Coca-Cola — Food/Beverages | Facebook," and "Starbucks | Facebook."

14. See Keller.

15. See Idle.

16. See Kendrick.

17. See Sullivan.

18. See Warner.

19. See Conti.

CHAPTER 7

1. See Ross.

2. See West, p. ix.

3. See Dick.

4. See above.

5. See Henderson, p. 230.

6. See Tang.

7. See Salamon, p. 96.

8. See Perednia, p. 312.

9. See Wang, S.

10. See Reid, p. 57–59.

11. See "Doctors' Use of Mobile Phone Apps Rising, Says Study."

12. See "Overweight and Obesity Statistics."

13. See Zundel.

14. See Scalvini.

15. See Ross.

16. See Wang, Hao.

17. See Mongan, p. 1510.

18. See Sabaté.

19. See Hobson.

20. See Hyman.

21. See West, p. x.

22. See Ross.

23. See Wang, S.

24. See Sullivan.

25. See "Diabetes and Pre-diabetes Statistics and Facts."

26. See Crane.

27. See "CT and MRI Scans Associated with Shorter Hospital Stays and Decreased Costs."

28. See Ross.

29. See Simon.

30. See Leiserson.

31. See Landro. "Hospitals See Multiple Benefits of EICUs."

32. See Landro. "The Picture of Health."

33. See above.

34. See Cavender.

35. See Ketabdar.

36. See "NFB — Blindness Statistics." Also cited in Azenkot.

37. See Azenkot.

38. See above.

39. See Roig-Franzia.

40. See Tung.

41. See "Heart Valve Replacement Surgery Abroad, Open Heart Valve Repair, Artificial Heart Valve Surgery."

42. See "The Cost of Knee Replacement Surgery."

43. See "Doctor."

44. See "Salary for Physician / Doctor, General Practice Jobs."

45. See Reid, p. 228.

46. According to the Deloitte Center for Health Solutions. See Johnson.

47. See Jehangir.

48. See Solow as cited in Zundel.

49. See "The Top 10 Causes of Death."

50. See above.

51. See "Hunger Stats."

52. See "The Top 10 Causes of Death."

53. See "Global Tuberculosis Control 2010."

54. See Green.

55. See "UNICEF — Progress for Children — How Many Are Underweight?"

56. See "UNICEF — Water, Sanitation and Hygiene — World Water Day 2005: 4,000 Children Die Each Day from a Lack of Safe Water."

57. See "UNICEF — Water, Sanitation and Hygiene — Statistics."

58. See "Neglected Tropical Diseases: Fast Facts."

59. See Hotez.

60. See "Energy — ICA."

61. See Msusa.

62. See Aker.

63. See Southwood, p. 38.

64. See O'Brien.

65. See "The World Factbook: GDP — Per Capita (PPP)."

66. See Auletta.

67. See Montgomery.

68. See Desai.

69. See Kahn.

70. See Green.

71. See Kaplan.

72. See Barclay.

73. See Shook.

74. See Reader, p. 245.

75. See Wernsdorfer, p. 105.

76. See Hotez.

77. See Nebehay.

78. See "Measles: Fact Sheet."

79. See Reader, p. 243.

80. See Blas.

81. See Hotez.

82. See Skolnik.

83. See Lederberg.

CHAPTER 8

1. See "Statistics Singapore — Time Series on Population."

2. See "Per Capita Personal Income."

3. See "Statistics Singapore — Time Series on Population."

4. See "GDP — per Capita (PPP) — Country Comparison."

5. See "Casino at Resorts World Sentosa, Singapore — Casino Levy."

6. See Glaeser, p. 228–229.

7. See "Statistics Singapore — Key Annual Indicators."

8. See Auguste.

9. See Patrinos, p. 22–23.

10. See Herbst.

11. See "Outcomes for College Graduates."

12. See Glaeser, p. 253.

13. See Hanushek.

14. See "Revenues and Expenditures for Public Elementary and Secondary Education — Selected Findings: Fiscal Year 2007."

15. See Rojas.

16. See OECD (2011).

17. See Schemo.

18. See Bosner, p. 4.

19. See Yazzie-Mintz.

20. See Wolfgang.

21. See "Facts and Figures: 2010–2011."

22. See D'Orio.

23. See Kim Hwa-young.

24. See Busch.

25. See D'Orio.

26. See Prince.

27. See Fischer.

28. See Voss.

29. See Prince.

30. See "Goddard Enters the World of Online Gaming."

31. See "All the (Synthetic) World's a Stage."

32. See *Edusim — 3D Virtual Worlds for the Classroom Interactive Whiteboard.*

33. See *Open Wonderland.*

34. See "Schools Lose Records; English Learners Pay."

35. See D'Orio.

36. See "CPS October 2009 — Detailed Tables — U.S. Census Bureau."

37. See "AIML: Artificial Intelligence Markup Language."

38. See "CPS October 2009 — Detailed Tables — U.S. Census Bureau."

39. See McCurry.

40. See "Revenues and Expenditures for Public Elementary and Secondary Education — Selected Findings: Fiscal Year 2007."

41. See Prince.

42. See Hinton.

43. See "College Prices Increase in Step with Inflation."

44. See "A History of College Inflation."

45. See Archibald, p. 25–28.

46. See "Princeton University | Fees & Payment Options."

47. See Burtless, p 118–119.

48. See Patel.

49. See Chaudhary.

50. See "Corruption in Education System in India."

51. See Luce, p. 77–78.

52. See Huebler.

53. See "Africa Education Watch: Good Governance Lessons for Primary Education."

54. See Najibullah.

55. See Shaw.

56. See MacWilliams.

57. See "Editorial: Civil and Political Rights in Schools."

58. See Alismail *et al*.

59. See Chen.

60. See *2012 World Hunger and Poverty Facts and Statistics*.

CHAPTER 9

1. See Apambire.

2. See Aker and Mbiti, "Mobile Phones and Economic Development in Africa."

3. See Aker and Mbiti, "Africa Calling: Can Mobile Phones Make a Miracle?"

4. See Lievrouw.

5. See "Electric Power Consumption (kWh per Capita) | Data | Table."

6. See Bridges.

7. See "Sanitation and Hygiene."

8. See O'Brien.

9. See "Global Mobile Statistics 2011: All Quality Mobile Marketing Research, Mobile Web Stats, Subscribers, Ad Revenue, Usage, Trends… | Mobi-Thinking."

10. See above.

11. See Jensen.

12. See above.

13. See Esipisu.

14. See Mittal.

15. See Kurata.

16. See Manjaro.

17. See Sullivan.

18. See "The World Factbook: Labor Force — By Occupation."

19. See "The World Factbook: GDP — Composition by Sector."

20. See Narula.
21. See Gandhi.
22. See Chaia.
23. See "Number of Credit Card Holders Slips to 18.3 Million."
24. See Nichter.
25. See Banerjee.
26. See Nichter.
27. See Banerjee.
28. See above.
29. See Yunus, p. 189–190.
30. See above, p. 51–52.
31. See Banerjee.
32. See Rosenberg.
33. See Jack.
34. See Rosenberg.
35. See above.
36. See "The Road to Hell Is Unpaved."
37. See Easterly, p. 111.
38. See Poisson, p. 3.
39. See Goel.
40. See Andersen.
41. See Grönlund, p. 18–19.
42. See "Mobile Services in Poor Countries: Not Just Talk."
43. See Faculty of Economics, University of Groningen.
44. See "China GDP Annual Growth Rate."
45. See Stepanek.
46. See van Meel.

CHAPTER 10

1. See Lawler.
2. See Zeitlin, p. 2.
3. See Lucas.
4. See "A History of American Agriculture: Farmers & the Land."
5. See Dimitri.
6. See "World FactBook."
7. See "Rural Labor and Education: Farm Labor."
8. See Hewitt.
9. See Sheppard.
10. See Manyika.

11. See Watters.

12. See "Building with Big Data."

13. See Manyika, p. 16.

14. See "World FactBook."

15. See "The American Workplace — The Shift to a Service Economy."

16. See Razin.

17. See Pélissié du Rausas.

18. See Manyika.

19. See Pélissié du Rausas.

20. See Lucas.

21. See *Cisco Visual Networking Index: Global Mobile Data Traffic Forecast Update, 2010–2015.*

22. See above.

23. See Warren, p. 196.

24. See Elliott.

REFERENCES

CHAPTER 1: THE WAVE

"Global Mobile Statistics 2011: All Quality Mobile Marketing Research, Mobile Web Stats, Subscribers, Ad Revenue, Usage, Trends . . . | MobiThinking." *Home | MobiThinking*. MobiThinking, July 2011. Web. 19 Oct. 2011. http://mobithinking.com/mobile -marketing-tools/latest-mobile-stats.

"Global Survey Shows Cell Phone Is 'Remote Control' for Life: 42 Percent of Americans 'Can't Live Without It' and Almost Half Sleep with It Nearby." *Global Market Research | Synovate*. Synovate: Research Reinvented, 2010. Web. 19 Oct. 2011. http://www.syno vate.com/news/article/2009/09/global-survey-shows-cell-phone-is-remote-control-for-life-42-of-americans-can-t-live-without-it-and-almost-half-sleep-with-it-nearby.html.

Golgowski, Nina. "Facebook IPO Valued at over \$100bn: Social Networking Giant 'to Go Public Next Spring' | Mail Online." *Mail Online*. Associated Newspapers Ltd, 30 Nov. 2011. Web. 16 Dec. 2011. http://www.dailymail.co.uk/news/article-2067634/Facebook -IPO-valued-100bn-Social-networking-giant-public-spring.html.

"Google Net Worth 2011." *Exploredia—Interesting News and Facts*. Exploredia, 1 Mar. 2011. Web. Sept. 2011. http://exploredia.com/google-net-worth-2011/.

Hasay, Kathleen. "ITunes, UTunes, We-all-Tunes and the Death of the Album Looms." *USC: InsightBusiness*. University of Southern California, 2008. Web. Nov. 2011. http://www.usc.edu/org/InsightBusiness/ib/articles/articlescontent/08_4_Katherine %20Hasay2.html.

"Sabre: The First Online Reservation System." *IBM 100*. IBM. Web. Nov. 2011. http://www.ibm.com/ibm100/us/en/icons/sabre/breakthroughs/.

Shah, Anup. "Poverty Facts and Stats—Global Issues." *Global Issues: Social, Political, Economic and Environmental Issues That Affect Us All—Global Issues*. Global Issues, 20 Sept. 2010. Web. 19 Oct. 2011. http://www.globalissues.org/article/26/poverty-facts -and-stats.

"Statistics | Facebook." *Facebook.com*. Facebook, 2011. Web. Nov. 2011. http://www.face book.com/press/info.php?statistics.

Trotsky, Leon. "A Letter to Albert Goldman." *In Defense of Marxism*. Marxist.org. 1942. Web. 19 Oct. 2011. http://www.marxists.org/archive/trotsky/works/pdf/defmarx.pdf.

CHAPTER 2: COMPUTERS

Ahonen, T. T. Communities Dominate Brands: Celebrating 30 Years of Mobile Phones, Thank You NTT of Japan. *Communities Dominate Brands: Business and Marketing Challenges for the 21st Century*. 13 Nov. 2009. Web. Nov. 2011. http://comm unities-dominate.blogs.com/brands/2009/11/celebrating-30-years-of-mobile-phones -thank-you-ntt-of-japan.html.

"BlackBerry Users Hit by Eight-hour Outage—CNN." *Featured Articles from CNN*. CNN, 23 Dec. 2009. Web. 12 Nov. 2011. http://articles.cnn.com/2009 -12-23/tech/blackberry.outage_1_blackberry-customers-blackberry-subscribers -blackberry-internet-service?_s=PM:TECH.

Chmielewski, Dawn, and Meg James, "Spending Pattern Spins Hollywood," *Los Angeles Times*, 18 Jan. 2011, p. B3.

Cortada, James W. *The Digital Hand*. Oxford: Oxford UP, 2004, p. 157.

Epstein, Zach. "IOS Market Share Balloons in October, Android Climbs to No. 2 Mobile OS." *BGR: The Three Biggest Letters In Tech*. BGR Media, LLC, 1 Nov. 2011. Web. 12 Dec. 2011. http://www.bgr.com/2011/11/01/ios-market-share-balloons-in -october-android-climbs-to-no-2-mobile-os/.

Fox, Steve, and Edward N. Albro, "No Second Coming: Apple's iPad Just a Big iPod Touch," *PCWorld*, 27 Jan. 2010. http://www.pcworld.com/article/187888/no _second_coming_apples_ipad_just_a_big_ipod_touch.html.

Gonsalves, Antone. "1 Billion Mobile Internet Devices Seen By 2013—Internet—Web Development—Informationweek." *InformationWeek | Business Technology News, Reviews and Blogs*. Information Week, 9 Dec. 2009. Web. 12 Dec. 2011. http://www.informationweek.com/news/internet/webdev/222001329.

Guglielmo, Connie. "Apple IPad Sales May Be as Much as Twice as Big as Some Analysts Estimated." *Bloomberg.com*. Bloomberg, 4 Apr. 2010. Web. Nov. 2011. http://www.bloomberg.com/news/2010-04-04/apple-ipad-debut-sales-may-be -beating-estimates-signaling-tablet-revival.html.

Hotz, Robert Lee. "In Art of Language, the Brain Matters—Los Angeles Times." Featured Articles from *The Los Angeles Times*. Los Angeles Times, 18 Oct. 1998. Web. 19 Oct. 2011. http://articles.latimes.com/1998/oct/18/news/mn-33918.

"IPhone4S Special Event 10/4/11—Scott Forstall Demos Siri Voice Recognition— YouTube." *YouTube—Broadcast Yourself*. TechnomiconMedia, 5 Oct. 2011. Web. 12 Nov. 2011. http://www.youtube.com/watch?v=EoVbACvhCDk.

Isaacson, Walter. *Steve Jobs*. New York: Simon & Schuster, 2011. Print.

Lashinsky, Adam. "The Decade of Steve Jobs, CEO of Apple—Nov. 5, 2009." *CNN Money*. CNN, 5 Nov. 2009. Web. 12 Nov. 2011. http://money.cnn.com/2009/11/04/technol ogy/steve_jobs_ceo_decade.fortune/index.htm.

Licklider, J.C.R. "Man-Computer Symbiosis," *IRE Transactions on Human Factors in Electronics*, volume HFE-1, March 1960, pp. 4–11.

Morris, Ian. *Why the West Rules—for Now: the Patterns of History, and What They Reveal about the Future*. New York: Farrar, Straus and Giroux, 2010. Print. p. 496.

Pelkey, J. (n.d.). Minicomputers, Distributed Data Processing and Microprocessors. *Entrepreneurial Capitalism and Innovation: A History of Computer Communications*

1968–1988. Web. Nov. 2011. http://www.historyofcomputercommunications.info /Book/5/5.1MinicomputersDistributedDataProcessingMicroprocessors.html.

Polsson, K. (2011, Nov. 1). Chronology of Personal Computers (1980). *Chronology of Personal Computers*. Web. Nov. 2011. http://pctimeline.info/comp1980.htm. .

Plowright, Catherine, Mathieu Lebeau, and Martine J. Perreault. "Relative Pattern Preferences by Bumblebees." *International Journal of Comparative Psychology*, 21 (2008), pp. 59–69.

"Sabre: The First Online Reservation System." *IBM 100*. IBM. Web. Nov. 2011. http://www.ibm.com/ibm100/us/en/icons/sabre/breakthroughs/.

Schlesinger, Henry R. *The Battery: How Portable Power Sparked a Technological Revolution*. Washington, D.C.: Smithsonian, 2010, p. 275.

Thomas, Manoj A., Richard T. Redmond, and Roland Weistroffer. "Moving to the Cloud: Transitioning From Client-Server To Service Architecture." *Journal of Service Science* 2.1 (2009).

Van Der Velden, Joanne, Ying Zheng, Blair Patullo, and David Macmillan. "PLoS ONE: Crayfish Recognize the Faces of Fight Opponents." PLoS ONE : *Accelerating the Publication of Peer-reviewed Science*. Web. 19 Oct. 2011. http://www.plosone.org/ article/info:doi/10.1371/journal.pone.0001695.

CHAPTER 3: PAPER

"AbitibiBowater Files for Bankruptcy Protection," *CBS News*, 16 Apr. 2009. http://www .cbc.ca/money/story/2009/04/16/mtl-abitibi-0416.html

Abramovitz, Janet N., and Ashley T. Mattoon. "Paper Cuts." *WorldWatch Institute*, paper #149, Dec. 1999. http://www.worldwatch.org/node/84.

"Annual (All Categories)." *NAA.org Homepage*. Web. 19 Oct. 2011. http://www.naa.org/ Trends-and-Numbers/Advertising-Expenditures/Annual-All-Categories.aspx

Battles, Matthew. *Library*. New York: W.W. Norton & Co., 2003.

Bawden, David, and Lyn Robinson, "A Distant Mirror?: The Internet and the Printing Press," *Aslib Proceedings,* vol. 52, no. 2, Feb. 2000–54, citing K. McGarry, The Changing Context of Information. 2nd ed. London: Library Association, 1993, p. 75.

Blodget, Henry. "Sulzberger Concedes: 'We Will Stop Printing *The New York Times* Sometime In The Future,'" 8 Sep 2010. http://www.businessinsider.com/sulzberger -we-will-stop-printing-the-new-york-times-2010-9.

Brown, Cynthia Stokes. *Big History: From the Big Bang to the Present*. New York: New Press, 2008. Print.

"Case: NFC in Oulu City Theatre." *SmartTouch*. VTT: Business from Technology, 2011. Web. 14 Feb. 2012. http://ttuki.vtt.fi/smarttouch/www/?info=case-theatre.

"Choose Cremation." *GreenYour: Funeral*. GreenYour. Web. Nov. 2011. http://www .greenyour.com/lifestyle/events/funeral/tips/choose-cremation.

Cody, Harold. "External Factors Squeezing Coated Freesheet Profits, Driving Supply/ Capacity Shifts and Price Changes." *PaperAge*, May/June, 2010, p. 14.

Crawford, J. (14 Oct. 2010). Inside Google Books: On the Future of Books. *Inside Google*

Books. Retrieved Nov. 2011. http://booksearch.blogspot.com/2010/10/on-future-of
-books.html

"Decline in Newsroom Jobs Slows." *ASNE.org*. American Society of News Editors, 4
Nov. 2010. Web. http://asne.org/article_view/articleid/763/decline-in-newsroom
-jobs-slows.aspx.

Dittmar, Jeremiah. "Information Technology and Economic Change: The Impact of the
Printing Press," *Monograph*, 14 Jan. 2010, p. 1.

Edmonds, Rick, Emily Guskin, and Tom Rosenstiel. "Newspapers: By the Numbers | State
of the Media." *The State of the News Media 2011*. Pew. Web. 20 Oct. 2011.
http://stateofthemedia.org/2011/newspapers-essay/data-page-6/.

Eisenstein, Elizabeth L. *The Printing Press as an Agent of Change*, vols. 1 and 2, New
York: Cambridge University Press, 1980.

"Employment by Industry, Occupation, and Percent Distribution, 2008 and Projected
2018." National Employment Matrix, Bureau of Labor Statistics, 2008. Web. 19
Oct. 2011. ftp://ftp.bls.gov/pub/special.requests/ep/ind-occ.matrix/ind_pdf/ind
_322000.pdf.

Febvre, Lucien, and Henri-Jean Martin. *The Coming of the Book: The Impact of Printing 1450–1800*, London: New Left Books, 1976. http://www.hrc.utexas.edu/educa
tor/modules/gutenberg/books/legacy/

Finkelstein, David, and Alistair McCleery. *An Introduction to Book History*. New York:
Routledge, 2005.

Flores, Marc. "Smartphone Owners Prefer Mobile for Breaking News." *Cell Phone News
and Reviews—IntoMobile*. IntoMobile, 7 Dec. 2010. Web. 19 Oct. 2011.
http://www.intomobile.com/2010/12/07/smartphone-mobile-breaking-news/.

"Frequent Questions | Paper Recycling | US EPA." US Environmental Protection Agency,
29 Nov. 2011. Web. Dec. 2011. http://www.epa.gov/osw/conserve/materials/
paper/faqs.htm.

"Frequently Asked Questions." *The Green Press Initiative—Working with Publishers, Industry Stakeholders and Authors to Create Paper-use Transformations That Will
Conserve Natural Resources and Preserve Endangered Forests. Green Press Initiative*. Web. 20 Oct. 2011. http://www.greenpressinitiative.org/about/faq.htm.

Fuchs, Robert. "The History and Biology of Parchment." Karger Gazette. Web. 19 Oct.
2011. http://www.karger.com/gazette/67/Fuchs/art_5.htm.

Füssel, Stephan. *Gutenberg and the Impact of Printing*. 2nd ed. Burlington: Ashgate,
2005. Print.

Goodrich, L. C. "Printing: Preliminary Report on a New Discovery." *Technology and Culture* 8.3 (1967): 376–378. Print.

Green Press Initiative. http://www.greenpressinitiative.org/about/faq.htm.

"Gutenberg's Legacy." *Harry Ransom Center*. University of Texas at Austin. Web. 19 Oct.
2011. http://www.hrc.utexas.edu/educator/modules/gutenberg/books/legacy/.

Halter, Reese. "Climate Change and Australia's Living Pinosaur." http://www.huffing
tonpost.com/dr-reese-halter/climate-change-and-austra_b_699438.html.

"High Operating Leverage Pressuring Newspaper Companies—Seeking Alpha." *Stock
Market News & Financial Analysis—Seeking Alpha*. Seeking Alpha, 5 June 2009.

Web. 19 Oct. 2011. http://seekingalpha.com/article/141644-high-operating-leverage-pressuring-newspaper-companies.

"Hippopotamuses: Fast Facts." *Wild Animal Facts*. National Geographic. Web. Nov. 2011. http://animals.nationalgeographic.com/animals/mammals/hippopotamus/.

Hurter, Arthur P., and Michael G. Van Buer. "The newspaper production/distribution problem," *Journal of Business Logistics*, 1996.

Jones, Sandra M. "Holiday Cards' Future Isn't Merry or Bright," *Los Angeles Times*, 9 Dec. 2010.

Kiehl, Jeffrey. "Lessons from Earth's Past." *Science* 331.6014 (2011): 158–159. Print.

Kirchhoff, Suzanne M. "The U.S. Newspaper Industry in Transition," Congressional Research Service, 9 Sept 2010, online, p. i.

Kramer, Samuel Noah. *History Begins at Sumer*. Philadelphia: University of Pennsylvania, 1981. Print.

Manguel, Alberto. *A History of Reading*. New York: Penguin, 1997.

Martin, Sam. "Paper Chase." *Ecology.com*. Web. 19 Oct. 2011. http://web.archive.org/web/20070619104819/http://www.ecology.com/feature-stories/paper-chase/index.html.

McCollan, Douglas, Julius J. Marke, Richard Sloane, and Linda M. Ryan, "The Changing Role of the Law Firm Library," in *Legal Research and Law Library Management*, Law Journal Press, 2006.

Meissner, Joern. *The Economics of E-Book Publishing*, Meissner Research Group, 17 Mar. 2010. http://www.meiss.com/blog/the-economics-of-e-book-publishing/.

North, Douglass C. and Robert Paul Thomas, *The Rise of the Western World: A New Economic History*, Cambridge, UK: Cambridge University Press, 1973.

"OECD Environmental Outlook (2001)." *Organisation for Economic Co-operation and Development*. Web. 19 Oct. 2011. http://www.oecd.org/document/26/0,3746,fr_2649_34283_1863386_1_1_1_1,00.html, p. 218.

Oregon Department of Environmental Quality. http://www.deq.state.or.us/lq/sw/cwrc/educate/facts.htm.

Pacey, Arnold. *Technology in World Civilization*. Cambridge, MA: MIT Press, 1991, p. 41.

Pettegree, Andrew. *The Book in the Renaissance*. New Haven, CT: Yale UP, 2010.

Pham, Alex. "Book Publishers See Their Role as Gatekeepers Shrink," *Los Angeles Times*, 23 Dec. 2010. http://www.latimes.com/business/la-fi-gatekeepers-20101226,0,7119214.story?track=rss&utm_source=feedburner&utm_medium=feed&utm_campaign=Feed%3A+latimes%2Fmostviewed+%28L.A.+Times+-+Most+Viewed+Stories%29.

"Project Gutenburg." New Canaan Library. Web. 19 Nov. 2011. http://newcanaanlibrary.org/articles/project-gutenberg.

Rich, Motoko. "Math of Publishing Meets the E-Book," *New York Times*, 28 Feb. 2010. http://www.nytimes.com/2010/03/01/business/media/01ebooks.html.

Robinson, Clark. "How Did We Get Here?" Newspaper Association of America, undated, http://www.naa.org/technews/tn971112/p6how.htm."

Sarno, David. "Libraries Reinvent Themselves as They Struggle to Remain Relevant in the Digital Age," *Los Angeles Times*, 12 Nov. 2010. http://www.latimes.com/business/la-fi-libraries-20101112,0,6514361.story.

Temple, Robert. *The Genius of China*, New York, Simon & Schuster, 1986.

"The New York Times Company Reports 2010 Third-Quarter Results." *New York Times Company: Investors Press Release*. New York Times Company, 19 Oct. 2010. Web. Nov. 2011. http://phx.corporate-ir.net/phoenix.zhtml?c=105317&p=irol-pressArticle&ID=1484239.

"Twenty Rugged Survivors in Dying Industries: The Commercial Printer: Suttle-Straus—BusinessWeek." *BusinessWeek Slide Shows and Multimedia*. BusinessWeek. Web. 20 Oct. 2011. http://images.businessweek.com/slideshows/20101116/twenty-rugged-survivors-in-dying-industries/slides/6.

"US Stats Show 9% Ebook Share, Grim News for Print." *EReport Digital Publishing Downunder*. EReport, 18 Oct. 2010. Web. 5 Mar. 2012. http://activitypress.com/blog/2010/10/18/us-stats-show-9-ebook-share-grim-news-for-print/.

"WAN—A Newspaper Timeline." *WAN-IFRA—Welcome to WAN-IFRA*. World Association of Newspapers, 2004. Web. 19 Oct. 2011. http://www.wan-press.org/article2822.html.

"WAN—Newspapers: A Brief History." *WAN-IFRA—Welcome to WAN-IFRA*. World Association of Newspapers, 2004. Web. 19 Oct. 2011. http://www.wan-press.org/article.php3?id_article=2821.

Wong, Wai. "Typesetting Chinese: A Personal Perspective." *TUGboat* 26.2 (2005): 111–14. Print.

Yu, Peter K. "Of Monks, Medieval Scribes, and Middlemen," *Michigan State Law Review* 1, (2006): 7.

Zacks Equity Research. "Publishing Industry Outlook—Nov. 2010." *Zacks.com*. Zacks Investment Research, 11 Nov. 2010. Web. 20 Oct. 2011. http://www.zacks.com/stock/news/43164/Publishing+Industry+Outlook.

CHAPTER 4: ENTERTAINMENT

Anton, Jason. "IPad Could Be Stealing Sales from Game Consoles." *Gamesradar.com*. Gamesradar, 9 July 2010. Web. 20 Oct. 2011. http://www.gamesradar.com/pc/pc/news/ipad-could-be-stealing-sales-from-game-consoles/a-20100709155042405084/g-2006032219817514003.

Arlotta, CJ. "CityVille Tops FarmVille's Highest Peak of Monthly Users—SocialTimes.com."*SocialTimes.com—Your Social Media Source*. Social Times, 3 Jan. 2011. Web. Dec. 2011. http://socialtimes.com/cityville-tops-farmvilles-highest-peak-of-monthly-users_b33272.

Burgan, Cristy. "Creating an SMS Opt-in Ecosystem for Mobile Advertising—Mobile Marketer—Columns." *Mobile Marketer—The News Leader in Mobile Marketing, Media and Commerce*. Mobile Marketer, 4 Aug. 2010. Web. 20 Oct. 2011. http://www.mobilemarketer.com/cms/opinion/columns/6988.html.

Chmielewski, Dawn C., and Meg James. "2010 Dvd Sales | For Hollywood, It Was a Tough 2010—Los Angeles Times." *Featured Articles from The Los Angeles Times*. Los Angeles Times, 18 Jan. 2011. Web. 20 Oct. 2011. http://articles.latimes.com/2011/jan/18/business/la-fi-ct-media-econ-20110118.

Clark, Don. "Qualcomm CEO Comes Clean About Mobile TV Miscues—Digits—WSJ." *WSJ Blogs—WSJ*. The Wall Street Journal, 1 Dec. 2010. Web. 20 Oct. 2011. http://blogs.wsj.com/digits/2010/12/01/qualcomm-ceo-comes-clean-about-mobile -tv-miscues/.

Clifford, Stephanie. "No Dice, No Money, No Cheating. Are You Sure This Is Monopoly?" *The New York Times: Business Day*. The New York Times, 15 Feb. 2011. Web. 20 Oct. 2011. http://www.nytimes.com/2011/02/16/business/16monopoly.html?page wanted=all.

"Console Wars | Seventh Generation | Worldwide Sales Figures." *Wikipedia, the Free Encyclopedia*. 2011. Web. Nov. 2011. http://en.wikipedia.org/wiki/Console_wars.

"Deloitte's 'State of the Media Democracy' Survey: TV Industry Embraces the Internet and Prospers." *Deloitte.com*. Deloitte, 1 Feb. 2011. Web. 20 Oct. 2011. http://www.deloitte.com/view/en_US/us/press/Press-Releases/dc69d100b 4ccd210VgnVCM2000001b56f00aRCRD.htm.

"Flickr.com Site Info." *Alexa the Web Information Company*. Flickr. Web. 18 Oct. 2011. http://www.alexa.com/siteinfo/Flickr.com.

Freeman, Mike. "Consumer Groups Seek to Block Former Flo TV Spectrum Sale." *SignOn San Diego*. San Diego News, 11 Mar. 2011. Web. Nov. 2011. http://www.signonsandiego.com/news/2011/mar/11/consumer-groups-seek-block -former-flo-tv-spectrum-/.

Global Mobile Statistics 2011: All Quality Mobile Marketing Research, Mobile Web Stats, Subscribers, Ad Revenue, Usage, Trends . . . | MobiThinking." *Home | MobiThinking*. Web. 15 Oct. 2011. http://mobithinking.com/mobile-marketing-tools/latest -mobile-stats.

Idvik, Lauren. "Twitter Set New Tweets Per Second Record During Super Bowl." *Social Media News and Web Tips "Mashable" The Social Media Guide*. Mashable, Inc, 9 Feb. 2011. Web. 18 Oct. 2011. http://mashable.com/2011/02/09/twitter-super-bowl -tweets/.

Ingram, Matthew. "Average Social Gamer Is a 43-Year-Old Woman." *GigaOM*. GigaOM, 17 Feb. 2010. Web. 18 Oct. 2011. http://gigaom.com/2010/02/17/average-social -gamer-is-a-43-year-old-woman/.

Krupa, Charlie. "Super Bowl 2011 Sets All-time TV Audience Record | NOLA.com." *New Orleans, LA Local News, Breaking News, Sports & Weather—NOLA.com*. New Orleans Net LLC, 7 Feb. 2011. Web. 18 Oct. 2011. http://www.nola.com/tv/ index.ssf/2011/02/super_bowl_2011_sets_all-time.html.

Mangalindan, J. P. "HTML5: Not Ready for Primetime, but Getting Very Close." *Fortune Tech: Technology Blogs, News and Analysis from Fortune Magazine*. CNN, 3 Dec. 2010. Web. 20 Oct. 2011. http://tech.fortune.cnn.com/2010/12/03/html5-not -ready-for-primetime-but-getting-very-close/.

Meyers, Justin. "Shocking FarmVille Facts! FarmVille World." FarmVille World. *Meet Other Players & Learn How To Make Your Farm The Best!* Oct. 2010. Web. 20 Oct. 2011. http://farmville.wonderhowto.com/blog/shocking-farmville-facts-0114622.

"Mobile Apps Convert a New Generation of Mass Market Casual Gamers—Gadgets & Tech, Life & Style—The Independent." *The Independent | News | UK and Worldwide*

News | Newspaper. Independent.co.uk, 23 Feb. 2011. Web. 20 Oct. 2011. http://www.independent.co.uk/life-style/gadgets-and-tech/mobile-apps-convert-a-new -generation-of-mass-market-casual-gamers-2223449.html?.

"Most Innovative Companies: Zynga." *FastCompany.com—Where Ideas and People Meet | Fast Company.* Fast Company, 2011. Web. 20 Oct. 2011. http://www.fastcompany.com/most-innovative-companies/2011/profile/zynga.php.

Netburn, Deborah. "Theaters Set aside Tweet Seats for Twitter Users—Latimes.com." *LA Times Business.* Los Angeles Times, 6 Dec. 2011. Web. Dec. 2011. http://latimesblogs.latimes.com/technology/2011/12/theaters-tweet-seats-twitter.html.

Nickinson, Phil. "Angry Birds Hits 500,000,000 Downloads | Android Central." *Android Central.* Mobile Nations, 2 Nov. 2011. Web. Dec. 2011. http://www.androidcentral.com/angry-birds-hits-500000000-downloads.

"Nielsen Says Video Game Penetration in U.S. TV Households Grew 18% During the Past Two Years." *PR Newswire: Press Release Distribution, Targeting, Monitoring and Marketing.* PR Newswire—United Business Media, 5 Mar. 2006. Web. Nov. 2011. http://www.prnewswire.com/news-releases/nielsen-says-video-game-penetration-in-us-tv-households-grew-18-during-the-past-two-years-51628587.html.

"Nielsen Wireless Survey, March 2010." *Mobile Momentum, How Consumer-Driven Competition Shapes & Defines the Modern U.S. Wireless Landscape,* 2010. www.mobilefuture.org.

O'Brien, Kevin J. "Mobile TV's Last Frontier—U.S. and Europe—NYTimes.com." *The New York Times: Business Day; Technology.* The New York Times, 30 May 2010. Web. 20 Oct. 2011. http://www.nytimes.com/2010/05/31/technology/31mobiletv.html.

Pew Research Center. "Internet & American Life Project, April 29–May 30, 2010 Tracking Survey." *Mobile Momentum, How Consumer-Driven Competition Shapes & Defines the Modern U.S. Wireless Landscape,* 2010. www.mobilefuture.org.

Purewal, Sarah Jacobsson. "15 Most Addictive IPhone Apps." *Reviews and News on Tech Products, Software and Downloads.* PC World, 14 Dec. 2010. Web. Nov. 2011. http://www.pcworld.com/article/213556/15_most_addictive_iphone_apps.html.

Rose, Frank. *The Art of Immersion: How the Digital Generation Is Remaking Hollywood, Madison Avenue, and the Way We Tell Stories.* New York: W.W. Norton & Co., 2011. Print.

Schneider, Michael. "TV Viewers' Average Age Hits 50." *Entertainment News, Film Reviews, Awards, Film Festivals, Box Office, Entertainment Industry Conferences.* Variety, 29 June 2008. Web. Nov. 2011. http://www.variety.com/article/VR1117988273.

Schonfeld, Erick. "Flurry: IPhone Games Are a $500 Million Industry in the U.S. and Taking Share." *TechCrunch.com.* TechCrunch, 22 Mar. 2010. Web. 20 Oct. 2011. http://techcrunch.com/2010/03/22/flurry-iphone-games-500-million/.

Selburn, Jordan. "Rising Media Tablet and Smartphone Sales Cut Demand for Single-Task Consumer Products." *ISuppli Press Release.* IHS ISuppli Market Research Group, 28 July 2011. Web. 13 Feb. 2012. http://www.isuppli.com/Home-and -Consumer-Electronics/News/Pages/Rising-Media-Tablet-and-Smartphone-Sales -Cut-Demand-for-Single-Task-Consumer-Products.aspx.

"Timeline of Computer History." *Computer History Museum*. Web. 20 Oct. 2011. http://www.computerhistory.org/timeline/?category=cmptr.

Upton, Chad. "More US Money Is Printed Than MONOPOLY Money « Broken Secrets." *Broken Secrets*. 4 Feb. 2011. Web. Dec. 2011. http://brokensecrets.com/2011/02/04/more-us-money-than-monopoly-money-printed-each-year/.

Valjalo, David. "Android Market Share on the Rise." *Video Game News, Reviews & Gaming Jobs*. Edge Magazine, 6 Dec. 2010. Web. Dec. 2011. http://origin.static.next-gen.biz/news/android-market-share-rise.

Wesch, Michael, "An Anthropological Introduction to YouTube," speech, Library of Congress, 23 June 2008. In Frank Rose, *The Art of Immersion*, New York, W.W. Norton & Co., 2011.

Wortham, Jenna. "Angry Birds, Flocking to Cellphones Everywhere—NYTimes.com." *The New York Times: Business Day Technology*. The New York Times, 11 Dec. 2010. Web. Dec. 2011. http://www.nytimes.com/2010/12/12/technology/12birds.html.

Zhang, Michael. "Flickr Passes the 5 Billion Photo Mark." *PetaPixel*. 18 Sept. 2010. Web. 20 Oct. 2011. http://www.petapixel.com/2010/09/18/flickr-passes-the-5-billion-photo-mark/.

CHAPTER 5: WALLET

"2008 Worldwide Credit Cards Usage Statistics and 2009 Trends." *Online Earnings Make Money & Online Tech Blog*. Online Earning, 2009. Web. Nov. 2011. http://onlinearnings.com/2009/06/18/2008-worldwide-credit-cards-usage-statistics-and-2009-trends/.

"Apple Events." *Steve Jobs Keynote*, 2 Mar. 2011. http://events.apple.com.edgesuite.net/1103pijanbdvaaj/event/index.html.

Ahern, Bill, Margaretta Blackwell, Jennifer Everett, Yolanda Ferguson, Margaret Furr, James Gregory, Jeff Kane, Ailsa Long, and Jim Strader. "Coin and Currency in the Casino Industry." *Richmondfed.org*. The Federal Reserve Bank of Richmond, 2005. Web. 20 Oct. 2011. http://www.richmondfed.org/banking/payments_services/understanding_payments/pdf/coin_and_currency_casino.pdf.

Bellens, Jan, Chris IP, and Anna McKinsey. "Developing a New Rural Payments System in China." *Ocnus.net*. Ocnus, 4 May 2007. Web. 20 Oct. 2011. http://www.ocnus.net/artman/publish/article_28854.shtml.

"Business Card History." *USRealtyCards.com . . . Your Source for Business Cards*. US Card Corporation. Web. 18 Oct. 2011. http://www.usrealtycards.com/business.card.history.html.

Calder, Lendol Glen. *Financing the American Dream a Cultural History of Consumer Credit*. Princeton, NJ: Princeton UP, 1999. Print.

Clark, Sarah. "Survey Shows Consumers Will Adopt Mobile Payments, Despite Security Fears." *Nfcworld.com*. SJB Research, 18 Feb. 2011. Web. 20 Oct. 2011. http://www.nearfieldcommunicationsworld.com/2011/02/18/36066/survey-shows-consumers-will-adopt-mobile-payments-despite-security-fears.

Cox, Brian L. "UPDATE: Police Seek Woman Who Allegedly Picked $5,000 from Store Owner's Pocket—Skokie News, Photos and Events—TribLocal.com." *Chicagoland*

News, Photos and Events—TribLocal.com. Chicago Tribune. Web. 18 Oct. 2011. http://triblocal.com/skokie/2011/01/13/police-see-woman-who-picked-5000-from -store-employees-pocket/.

Dukes, Anthony J. "Managerial Economics / Ch9 Pricing: Examples." *Managerial Economics / FrontPage*. PBworks, 2008. Web. 18 Oct. 2011. http://manecon.pbworks .com/w/page/20240341/Ch9-Pricing:-Examples.

Evans, David S., and Richard Schmalensee. *Paying with Plastic the Digital Revolution in Buying and Borrowing*. Cambridge, Mass.: MIT, 2005. Print.

Foley, Linda, Jay Foley, and Karen Barney. "Identity Theft: The Aftermath 2009." Identity Theft Resource Center, 2009. Web. 18 Oct. 2011. http://www.idtheftcenter.org/ artman2/uploads/1/Aftermath_2009_20100520.pdf.

Frellick, Marcia. "What Does It Cost to Get a Credit Card in Your Pocket?" *Credit Cards.com*. Credit Cards.com. 26 Aug. 2010. Web. Nov. 2011. http://www.credit cards.com/credit-card-news/cost-getting-credit-card-in-your-pocket-1276.php.

"Fun Facts About Money." *Frbsf.org*. The Federal Reserve Bank of San Francisco, 2011. Web. 20 Oct. 2011. http://www.frbsf.org/federalreserve/money/funfacts.html.

Gilfoyle, Timothy. "Street-rats and Gutter-snipes: Child Pickpockets and Street Culture in New York City, 1850–1900 | Journal of Social History | Find Articles." *Findarticles.com*. CBS, 2004. Web. 20 Oct. 2011. http://findarticles.com/p/articles /mi_m2005/is_4_37/ai_n6137400.

"Global Mobile Statistics 2011: All Quality Mobile Marketing Research, Mobile Web Stats, Subscribers, Ad Revenue, Usage, Trends . . . | MobiThinking." *Home | MobiThinking*. Web. 15 Oct. 2011. http://mobithinking.com/mobile-marketing-tools/ latest-mobile-stats

Heaton, Paul. "Hidden in Plain Sight: What Cost-of-Crime Research Can Tell Us About Investing in Police | RAND." *Rand.org*. Rand Corporation, 9 May 2011. Web. 20 Oct. 2011. http://www.rand.org/pubs/occasional_papers/OP279.html.

Heatwole, Anneryan. "The Mobile Minute: 90% of the World Has Access to Mobile Networks, Mobile Banking in the Philippines, and More." *MobileActive.org | A Global Network of People Using Mobile Technology for Social Impact*. MobileActive.org. 28 Oct. 2010. Web. Nov. 2011. http://mobileactive.org/mobile-minute-10-28-2010.

"In the Face of Danger: Facial Recognition and the Limits of Privacy Law," 120 *Harvard Law Review* 1870 (2007): 1875–1876.

Jimenez, Alberto, and Prasanna Vanguri. "Cash Replacement through Mobile Money in Emerging Markets: The FISA Approach." IBM Global Business Services, July 2010. Web. 20 Oct. 2011. http://www-935.ibm.com/services/us/gbs/pdf/GBW03122-USEN -02.pdf.

Keiningham, Timothy L. *Loyalty Myths: Hyped Strategies That Will Put You out of Business—and Proven Tactics That Really Work*. Hoboken, NJ: John Wiley & Sons, 2006. Print.

Karmin, Craig. *Biography of the Dollar: How the Mighty Buck Conquered the World and Why It's under Siege*. New York: Crown Business, 2008. Print.

Landrum, Shane. "Undocumented Citizens: The Crisis of U.S. Birth Certificates, 1940– 1945." *Cliotropic.org*. 2011. Web. 20 Oct. 2011. http://cliotropic.org/blog/talks/ undocumented-citizens-aha-2010.

"Making a Dollar Bill Now Costs 9.6 Cents." *Opportunist Magazine*. Ed. Phil Robertson. Opportunist Magazine, 26 May 2011. Web. Dec. 2011. http://opportunist magazine.com/making-a-dollar-bill-now-costs-9-6-cents-name/.

Martin, Preston, and Lita Epstein. *The Complete Idiot's Guide to the Federal Reserve*. Indianapolis, IN: Alpha, 2003. Print.

"Moving Assembly Line at Ford." 7 Oct. 2011. http://www.history.com/this-day-in -history/moving-assembly-line-at-ford.

"Official Identity Theft Statistics | SPENDonLIFE." *Check Your Free Credit Score, Get All 3 Scores Online Instantly*. SpendOnLife.com. Web. 18 Oct. 2011. http://www .spendonlife.com/guide/identity-theft-statistics.

Petruno, Tom. "Credit Card Debt Rises for First Time in Two Years—Latimes.com." *Blogs—Latimes.com*. Los Angeles Times, 7 Feb. 2011. Web. 20 Oct. 2011. http:// latimesblogs.latimes.com/money_co/2011/02/credit-card-debt-balances-rise -consumer-spending-fed-economy.html.

Preson, Marjorie. "Taking the Lead." *Tribal Government Gaming*. 13 Jan. 2010. Web. 20 Oct. 2011. http://tribalgovernmentgaming.com/issue/tribal-government-gaming -2008/article/taking-the-lead.

Shin, Laura. "Making Credit Cards Landfill-Friendly—NYTimes.com." *Energy and Environment—Green Blog—NYTimes.com*. The New York Times, 23 Feb. 2009. Web. 20 Oct. 2011. http://green.blogs.nytimes.com/2009/02/23/making-credit-cards-land fill-friendly/.

Sims, Calvin. "In Recycling of Greenbacks, New Meaning for Old Money." *The New York Times—Breaking News, World News & Multimedia*. The New York Times, 22 May 1994. Web. Nov. 2011. http://www.nytimes.com/1994/05/22/us/in-recycling -of-greenbacks-new-meaning-for-old-money.html?pagewanted=all.

Smith, David. *The Future of the Internet–'You Ain't Seen Nothin' Yet.'* IoD Big Picture, Quarter 3, 2011. Print.

"The Lowdown On Customer Loyalty Programs—Forbes.com." *Information for the World's Business Leaders—Forbes.com*. Forbes.com, 01 Feb. 2007. Web. 18 Oct. 2011. http://www.forbes.com/2007/01/02/frequent-flyer-miles-ent-sales-cx_kw_0102 whartonloyalty.html.

"The United States Mint · About Us." *The United States Mint*. Department of the Treasury, 2010. Web. 18 Oct. 2011. http://www.usmint.gov/about_the_mint/coin _production/index.cfm?action=production_figures.

Truman, Jennifer, and Michael R. Rand. "Criminal Victimization, 2009." Washington, D.C.: U.S. Department of Justice, October 2010.

Tuttle, Brad. "2010: Huge Year for Coupons, Especially in the South | Moneyland | TIME.com." *Moneyland | Financial Insights from Your Wallet to Wall Street | TIME.com*. Time Magazine, 24 Jan. 2011. Web. 20 Oct. 2011. http://money.blogs .time.com/2011/01/24/2010-huge-year-for-coupons-especially-in-the-south/.

"United States Circulating Coinage Intrinsic Value Table." *Current Melt Value of Coins— How Much Is Your Coin Worth?* Coinflation, 18 Oct. 2011. Web. 18 Oct. 2011. http://www.coinflation.com/.

U.S. Bureau of Engraving and Printing—"Annual Production Figures." *U.S. Bureau of Engraving and Printing—Home*. Bureau of Engraving and Printing: U.S. Department of Treasury. Web. 15 Oct. 2011. http://www.moneyfactory.gov/uscurrency/annual productionfigures.html.

Valdes-Dapena, Peter. "Auto Theft Rates Lowest since 1967—Jun. 21, 2011." *CNNMoney— Business, Financial and Personal Finance News*. CNN, 21 June 2011. Web. 14 Feb. 2012. http://money.cnn.com/2011/06/21/autos/auto_theft_rate_nicb/index.htm.

Valsic, Bill. "Via Zipcar Ford Seeks Young Fans." *The New York Times*. 31 Aug. 2011. Print.

Vavra, Terry, Lerzan Aksoy, and Henri Wallard. "Abstract." Introduction. *Loyalty Myths*. By Timothy Keiningham. Hoboken, NJ: John Wiley & Sons, 2005. Ipsosloyalty, 30 Nov. 2005. Web. Nov. 2011. http://www.ipsos.com/loyalty/sites/ipsos.com.loyalty/files/Ipsos_Loyalty_Myth_31_Excerpt_0.pdf.

Voorhees, Don. *The Essential Book of Useless Information: The Most Unimportant Things You'll Never Need to Know*. New York: Perigee, 2010. Print.

Wang, Jim. "50 Fun Facts About Cold Hard Cash." *Bargaineering*. Bargaineering.com, 29 Jan. 2008. Web. Nov. 2011. http://www.bargaineering.com/articles/50-fun-facts -about-cold-hard-cash.html.

Weatherford, J. McIver. *The History of Money: From Sandstone to Cyberspace*. New York: Crown, 1997. Print.

Woolsey, Ben, and Matt Schulz. "Credit Card Statistics, Industry Facts, Debt Statistics." *CreditCards.com*. Credit Cards.com, 2011. Web. 20 Oct. 2011. http://www.credit cards.com/credit-card-news/credit-card-industry-facts-personal-debt-statistics-1276.php. In "The Survey of Consumer Payment Choice," Federal Reserve Bank of Boston, January 2010.

Yule, Henry, ed. *The Book of Ser Marco Polo, the Venetian: Concerning the Kingdoms and Marvels of the East*. Vol. 1. Cambridge: Cambridge UP, 2010. Print.

Zetter, Kim. "Feds Swoop In on Nationwide Pickpocket, I.D. Theft Ring | Threat Level | Wired.com." *Wired.com*. Wired, 10 June 2009. Web. 18 Oct. 2011. http://www.wired .com/threatlevel/2009/06/pickpockets/.

"Zipcar to Replace City Vehicles in New Deal Saving $400K." *Chicago News and Weather | Fox Chicago News*. Fox Chicago News, 4 Mar. 2011. Web. Oct. 2011. http://www.myfoxchicago.com/dpp/news/metro/zipcar-city-chicago-vehicles-deal -saving-budget-20110304.

CHAPTER 6: SOCIAL NETWORKS

Adegoke, Yinka. "How News Corp Got Lost in MySpace." *The Globe and Mail: Tech News*. The Globe and Mail, Inc., 7 Apr. 2011. Web. 8 Nov. 2011. http://m.theglobe andmail.com/news/technology/tech-news/how-news-corp-got-lost-in-myspace/article1974889/?service=mobile.

Calvo-Armengol, Antoni, and Matthew O. Jackson. "The Effects of Social Networks on Employment and Inequality." *American Economic Review*, v. 94, 3 (June 2004): 426–454.

Carter, Michelle et al., "What Cell Phones Mean in Young People's Daily Lives and Social Interactions," *Proceedings of the Southern Association for Information Systems Conference*, Atlanta, GA, 25–26 Mar. 2011.

"Cell Phones Key to Teens' Social Lives, 47% Can Text with Eyes Closed." *Marketingcharts.com*. HarrisInteractive, 23 Sept. 2008. Web. 22 Oct. 2011. http://www.marketingcharts.com/interactive/cell-phones-key-to-teens-social-lives-47-can-text-with-eyes-closed-6126.

Chafkin, Max. "How to Kill a Great Idea!" *Inc.com*. Inc Magazine, 1 June 2007. Web. 22 Oct. 2011. http://www.inc.com/magazine/ 20070601/features-how-to-kill-a-great-idea.html.

Conti, Mauro, Arbnor Hasani, and Bruno Crispo. "Virtual Private Social Networks," *CODASPY'11*, 21–23 Feb. 2011.

El-Heni, Zied. "Reporter's Notebook: Tunisia." *The Global Integrity Report*. Global Integrity, 2008. Web. 8 Nov. 2011. http://report.globalintegrity.org/Tunisia/2008/notebook.

"Eminem | Facebook." *Welcome to Facebook*. Eminem, 13 Dec. 2011. Web. 13 Dec. 2011. http://www.facebook.com/eminem?ref=.

Ghelawat, Sunil, Kenneth Radke, and Margot Brereton. "Interaction, Privacy and Profiling Considerations in Local Mobile Social Software: a Prototype Agile Ride Share System." *ACM Digital Library*. ACM, 22 Nov. 2010. Web. 22 Oct. 2011. http://portal.acm.org/dl.cfm.

Hölldobler, Bert, Edward O. Wilson, and Margaret Cecile Nelson. *The Superorganism: the Beauty, Elegance, and Strangeness of Insect Societies*. New York: W.W. Norton, 2009. Print.

Idle, Nadia, and Alex Nunns. "Tahrir Square Tweet by Tweet | World News | The Guardian." *Guardian.co.uk*. The Guardian, 14 Apr. 2011. Web. 22 Oct. 2011. http://www.guardian.co.uk/world/2011/apr/14/tahrir-square-tweet-egyptian-uprising.

Keller, Jared. "When Campaigns Manipulate Social Media—Jared Keller—Politics—The Atlantic." *The Atlantic—News and Analysis on Politics, Business, Culture, Technology, National, International, and Life–TheAtlantic.com*. The Atlantic, 10 Nov. 2010. Web. 22 Oct. 2011. http://www.theatlantic.com/politics/archive/2010/11/when-campaigns-manipulate-social-media/66351.

Kendrick, James. "Trapster Hacked: 10 Million Mobile Users Potentially Affected | ZDNet." *Technology News, Analysis, Comments and Product Reviews for IT Professionals*. ZD Net, 20 Jan. 2011. Web. Nov. 2011. http://www.zdnet.com/blog/mobile-news/trapster-hacked-10-million-mobile-users-potentially-affected/608.

Lenhart, Amanda et al., *Teens and Mobile Phones*. Washington, DC: Pew Internet & American Life Project, 20 Apr. 2010.

Nakashima, Ryan. "MySpace Layoffs: 47 Percent Of Staff Fired." *Breaking News and Opinion on The Huffington Post*. Huffington Post, 11 Jan. 2011. Web. 22 Oct. 2011. http://www.huffingtonpost.com/2011/01/11/myspace-layoffs-2011_n_807421.html.

Saylor, Michael. "CEO Keynote Address." MicroStrategy World Conference. Hyatt Regency Miami. Miami, FL. 23 Jan. 2012. Keynote Address.

Schonfeld, Erick. "Costolo: Twitter Now Has 190 Million Users Tweeting 65 Million Times A Day." *Tech Crunch*. AOL, 8 June 2010. Web. Nov. 2011. http://techcrunch.com/2010/06/08/twitter-190-million-users/.

Sullivan, Bob. "Craigslist 'Robberies by Appointment' Turn Violent." *Msnbc.com*. MSNBC, 7 Jan. 2011. Web. 22 Oct. 2011. http://redtape.msnbc.msn.com/_news/2011/01/07/6345499-craigslist-robberies-by-appointment-turn-violent.

Vermeij, Geerat J. *Nature: an Economic History*. Princeton: Princeton UP, 2004. Print.

Warner, Bill. "Craigslist Killer Philip Markoff Found Dead in Boston Jail Cell, Private Investigator Bill Warner Profiled His Type in National Article 'The Dangerous Side of Online Dating'. Bill Warner Sarasota Private Investigator Catch Cheaters 941-926 -1926." *Pibillwarner.wordpress.com*. Bill Warner Sarasota Private Investigator Catch Cheaters 941-926-1926, 15 Aug. 2010. Web. 22 Oct. 2011. http://pibillwarner.word press.com/2010/08/15/craigslist-killer-philip-markoff-found-dead-in-boston-jail-cell-private-investigator-bill-warner-profiled-his-type-in-national-article-the-dangerous -side-of-online-dating/.

CHAPTER 7: MEDICINE

Aker, Jenny C., and Isaac M. Mbiti. "Mobile Phones and Economic Development in Africa." *Journal of Economic Perspectives* 24.3 (2010): 207–232. Print.

Auletta, Ken. "The Dictator Index." *The New Yorker*. 7 Mar. 2011. Web. http://www.new yorker.com/reporting/2011/03/07/110307fa_fact_auletta.

Azenkot, Shiri, et al. "Enhancing Independence and Safety for Blind and Deaf-Blind Public Transit Riders." *CHI 2011* (2011). *Cs.washington.edu*. University of Washington, 2011. Web. http://www.cs.washington.edu/homes/shiri/papers/azenkot_chi 2011.pdf.

Barclay, Eliza. "Text Messages Could Hasten Tuberculosis Drug Compliance." *The Lancet* 373.9657 (2009): 15–16. Print.

Blas, B. L., I. L. Lipayon, L. C. Tormis, L. A. Portillo, M. Hayashi, and H. Matsuda. "An Attempt to Study the Economic Loss Arising from Schistosoma Japonicum Infection and the Benefits Derived from Treatment." *Southeast Asian J Trop Med Public Health* 37.1 (2006): 26–32. *PubMed.gov*. Web. http://www.ncbi.nlm.nih.gov/pubmed/16771209.

Cavender, Anna, Richard E. Ladner, and Eve A. Riskin. "MobileASL: Intelligibility of Sign Language Video as Constrained by Mobile Phone Technology." *ASSETS '06* (22–25 Oct. 2006). Print.

Crane, Misti. "Device in Artery Helps Track Blood Pressure in Heart Patients | The Columbus Dispatch." *Dispatch.com*. The Columbus Dispatch, 9 Feb. 2011. Web. 22 Oct. 2011. http://www.dispatch.com/live/content/local_news/stories/2011/02/09/device-in-artery-helps-track-blood-pressure-in-heart-patients.html?sid=101.

"CT and MRI Scans Associated with Shorter Hospital Stays and Decreased Costs." *Science Daily: News & Articles in Science, Health, Environment & Technology*.

ScienceDaily, 1 Apr. 2010. Web. 22 Oct. 2011. http://www.sciencedaily.com/releases/2010/04/100401085352.htm.

Desai, Lisa. "Cell Phones Save Lives in Rwandan Villages—CNN.com." *CNN.com International—Breaking, World, Business, Sports, Entertainment and Video News.* CNN, 28 July 2010. Web. 23 Oct. 2011. http://edition.cnn.com/2010/WORLD/africa/07/28/Rwanda.phones.pregnant.women/index.html?eref=edition_africa.

"Diabetes and Pre-diabetes Statistics and Facts." *National Diabetes Education Program.* National Institute of Diabetes and Digestive and Kidney Diseases. Web. 8 Nov. 2011. http://ndep.nih.gov/diabetes-facts/index.aspx.

Dick, Richard S., and Elaine B. Steen. *The Computer-based Patient Record: An Essential Technology for Health Care.* Washington, D.C.: National Academy, 1991. Print.

"Doctor." *Www.bls.gov.* U.S. Bureau of Labor Statistics, July 2011. Web. 23 Oct. 2011. http://www.bls.gov/k12/help06.htm.

"Doctors' Use of Mobile Phone Apps Rising, Says Study." *PharmaTimes Online.* PharmaTimes, 5 Jan. 2011. Web. 22 Oct. 2011. http://www.pharmatimes.com/article/11-01-05/Doctors_use_of_mobile_phone_apps_rising_says_study.aspx.

"Energy—ICA." *Icafrica.org.* Infrastructure Consortium for Africa, 2009. Web. 8 Nov. 2011. http://www.icafrica.org/en/infrastructure-issues/energy/.

Eric Topol: The Wireless Future of Medicine | *Video on TED.com.* Perf. Eric Topol. *TED: Ideas worth Spreading.* TEDMED, Feb. 2010. Web. 22 Oct. 2011. http://www.ted.com/talks/eric_topol_the_wireless_future_of_medicine.html.

"Global Tuberculosis Control 2010." *www.who.int.* World Health Organization, 2009. Web. 8 Nov. 2011. http://www.who.int/tb/publications/global_report/2010/en/index.html.

Green, Edward C. *Broken Promises: How the AIDS Establishment Has Betrayed the Developing World.* Sausalito, CA: PoliPoint, 2011. Print.

Henderson, James W. *Health Economics & Policy.* Mason, OH: South-Western, 2009. Print.

"Heart Valve Replacement Surgery Abroad, Open Heart Valve Repair, Artificial Heart Valve Surgery." *Medical Tourism.* PlacidWay, LLC. Web. 8 Nov. 2011. http://www.placidway.com/subtreatment-detail/treatment,13,subtreatment,47.html/Heart-Valve-Replacement-Surgery-Treatment-Abroad.

Hobson, Katherine. "How Can You Help the Medicine Go Down?" *The Wall Street Journal: Innovations in Health Care.* The Wall Street Journal, 28 Mar. 2011. Web. 22 Oct. 2011. http://online.wsj.com/article/SB10001424052748703386704576186370263245588.html.

Hotez, Peter J., Alan Fenwick, Lorenzo Savioli, and David H. Molyneux. "Rescuing the Bottom Billion through Control of Neglected Tropical Diseases." *The Lancet.* 373.9674 (2009): 1570–1575. Print.

"Hunger Stats." *Wfp.org.* World Food Programme, 2007. Web. 8 Nov. 2011. http://www.wfp.org/hunger/stats.

Hyman, Paul. "Physicians See Mobile Phones as Tools to Aid Non-Compliant Patients | News." *Communications of the ACM.* ACM, 21 Sept. 2010. Web. 22 Oct. 2011.

http://cacm.acm.org/news/99116-physicians-see-mobile-phones-as-tools-to-aid
-non-compliant-patients/fulltext.

Jehangir, Bakshi, and Asifa Nazir. "Telemedicine—Doctors Without Borders." *JK—Practitioner* 10.2 (2003): 158–159. Web. 24 Oct. 2011. http://medind.nic.in/jab/t03
/i2/jabt03i2p158g.pdf.

Johnson, Linda A. "Americans Look Abroad to Save on Health Care / Medical Tourism
Could Jump Tenfold in next Decade." *Sfgate.com*. San Francisco Chronicle, 3 Aug.
2008. Web. 23 Oct. 2011. http://www.sfgate.com/cgi-bin/article.cgi?f=/c/a/2008/08/
03/BUGA121GPF.DTL.

Kahn, J. G., J. S. Yang, and J. S. Kahn. "'Mobile' Health Needs and Opportunities in
Developing Countries." *Health Affairs* 29.2 (2010): 254–261. Print.

Kaplan, Melanie D. G. "Bill Gates: Mobile Health Technology Will save Lives, Help Overpopulation | SmartPlanet." *SmartPlanet—Innovative Ideas That Impact Your World*.
Smartplanet, 10 Nov. 2010. Web. 23 Oct. 2011. http://www.smartplanet.com/
blog/pure-genius/bill-gates-mobile-health-technology-will-save-lives-help-overpop
ulation/4908.

Ketabdar, Hamed, and Tim Polzehl. "Tactile and Visual Alerts for Deaf People by Mobile
Phones." *ASSETS '09* (2009). Print.

Landro, Laura. "Hospitals See Multiple Benefits of EICUs." *The Wall Street Journal:
Health Care*. The Wall Street Journal, 27 Oct. 2009. Web. 08 Nov. 2011. http://
online.wsj.com/article/SB10001424052970204488304574428960127233136.html.

Landro, Laura. "The Picture of Health." *The Wall Street Journal: Health Care*. The Wall
Street Journal, 27 Oct. 2009. Web. 22 Oct. 2011. http://online.wsj.com/article/
SB10001424052970204488304574428960127233136.html.

Lederberg, Joshua. "Infectious History." *Science* 14 (2000): 287–93. Print.

Leiserson, Mark. "The Future of Medical Imaging." *Tuftscope* Spring 9.11 (2010):
17–18. Print.

"Measles: Fact Sheet." *Www.who.int*. World Health Organization, Oct. 2011. Web. 08
Nov. 2011. http://www.who.int/mediacentre/factsheets/fs286/en/.

Mongan, James J., Timothy G. Ferris, and Thomas H. Lee. "Options for Slowing the
Growth of Health Care Costs." *New England Journal of Medicine* 358.14 (2008):
1509–1514. Print.

Montgomery, Maggie. *Sustaining Trachoma Control and Elimination*. World Health
Organization, 2006. Print.

Msusa, Sheen. "Rethinking Africas Roads Network Challenges." *ModernGhana.com*.
Modern Ghana Media Group, 4 Dec. 2007. Web. 08 Nov. 2011. http://www
.modernghana.com/news/148282/1/rethinking-africas-roads-network-challenges.html.

Nebehay, Stephanie. "Deadly Measles Outbreaks Threaten Africa Gains: WHO| Reuters."
Reuters.com. Reuters, 21 May 2010. Web. 23 Oct. 2011. http://www.reuters.com/
article/2010/05/21/us-measles-idUSTRE64K2Y220100521.

"Neglected Tropical Diseases: Fast Facts." *CDC.gov*. Centers for Disease Control and
Prevention, 6 June 2011. Web. 8 Nov. 2011. http://www.cdc.gov/globalhealth/
ntd/fastfacts.html.

"NFB—Blindness Statistics." *NFB—Home*. National Federation of the Blind, 2011. Web.
23 Oct. 2011. http://www.nfb.org/nfb/blindness_statistics.asp.

O'Brien, Kevin J., "Nokia Taking a Rural Road to Growth." *New York Times.* 1 Nov 2010

"Overweight and Obesity Statistics." *Weight-Control Information Network.* National Institute of Diabetes and Digestive and Kidney Diseases, 5 July 2011. Web. 8 Nov. 2011. http://win.niddk.nih.gov/statistics/.

Perednia, Douglas A. *Overhauling America's Healthcare Machine: Stop the Bleeding and Save Trillions.* Upper Saddle River, NJ: FT, 2011. Print.

Reader, John. *Africa: a Biography of the Continent.* New York: Vintage, 1999. Print.

Reid, T. R. *The Healing of America: a Global Quest for Better, Cheaper, and Fairer Health Care.* New York: Penguin, 2009. Print.

Roig-Franzia, Manuel. "Discount Dentistry, South of The Border." *Washingtonpost.com.* The Washington Post, 18 June 2007. Web. 23 Oct. 2011. http://www.washington post.com/wp-dyn/content/article/2007/06/17/AR2007061701297.html.

Ross, Philip. "Managing Care Through the Air." *IEEE Spectrum.* December 2004.

Sabaté, Eduardo. *Adherence to Long-term Therapies: Evidence for Action.* Geneva: World Health Organization, 2003. Web. 20 Oct. 2011. http://www.who.int/chp/know ledge/publications/adherence_full_report.pdf.

Salamon, Julie. *Hospital: Man, Woman, Birth, Death, Infinity, plus Red Tape, Bad Behavior, Money, God, and Diversity on Steroids.* New York: Penguin, 2008. Print.

"Salary for Physician / Doctor, General Practice Jobs." *Www.payscale.com.* PayScale, 17 Oct. 2011. Web. 23 Oct. 2011. http://www.payscale.com/research/IN/Job= Physician_/_Doctor,_General_Practice/Salary.

Scalvini, S., et al. "Telemedicine: A New Frontier for Effective Healthcare Services." *PubMed Central* 61.4 (2004): 226–233. Print.

Shook, F. M., M.D. "Prophylaxis of Some Tropical Infections." *New York State Journal of Medicine* 12.11 (1912): 630–633. Print.

Simon, Stephanie. "Medicine on the Move: Mobile Devices Help Improve Treatment." *Wall Street Journal Online.* Wall Street Journal, 28 Mar. 2011. Web. http:// online.wsj.com/article/SB10001424052748703559604576174842490398186.htm.

Skolnik, Richard, and Ambareen Ahmed. "Ending the Neglect of Neglected Tropical Diseases—Population Reference Bureau." *Population Reference Bureau.* Population Reference Bureau, Feb. 2010. Web. 23 Oct. 2011. http://www.prb.org/Reports/ 2010/neglectedtropicaldiseases.aspx.

Southwood, Russell. *Less Walk, More Talk: How Celtel and the Mobile Phone Changed Africa.* Chichester, West Sussex, England: John Wiley, 2008. Print.

Sullivan, Thomas. "Journal of Clinical Hypertension: CME Learners Provide Better Patient Care." *Policy and Medicine.* CME, 6 Jan. 2011. Web. 8 Nov. 2011. http://www.policymed.com/2011/01/journal-of-clinical-hypertension-cme-learners -provide-better-patient-care.html.

Tang, MD, Paul C., Danielle Fafchamps, and Edward H. Shortliffe, MD. "Traditional Medical Records as a Source of Clinical Data in the Outpatient Setting." *PubMed Central* (1994): 575–579. Print.

"The Cost of Knee Replacement Surgery." *Knee Replacement Cost.com.* Formosa Medical Travel. Web. 8 Nov. 2011. http://www.kneereplacementcost.com/.

"The Top 10 Causes of Death." *Www.who.int*. World Health Organization. Web. 23 Oct. 2011. http://www.who.int/mediacentre/factsheets/fs310/en/index2.html.

"The World Factbook: GDP—Per Capita (PPP)." *Cia.gov*. Central Intelligence Agency. Web. 8 Nov. 2011. https://www.cia.gov/library/publications/the-world-factbook/rankorder/2004rank.html.

Tung, Sarah. "Taiwan's Cosmetic-Surgery Sector Looks to Mainland China—TIME."*Time.com*. Time Magazine, 16 July 2010. Web. 23 Oct. 2011. http://www.time.com/time/world/article/0,8599,2004023,00.html.

"UNICEF—Progress for Children—How Many Are Underweight?" *UNICEF—UNICEF Home*. UNICEF, 4 May 2006. Web. 23 Oct. 2011. http://www.unicef.org/progressforchildren/2006n4/index_howmany.html.

"UNICEF—Water, Sanitation and Hygiene—Statistics." *UNICEF—UNICEF Home*. UNICEF, 27 Apr. 2010. Web. 14 Feb. 2012. http://www.unicef.org/wash/index_statistics.html.

"UNICEF—Water, Sanitation and Hygiene—World Water Day 2005: 4,000 Children Die Each Day from a Lack of Safe Water." *UNICEF—UNICEF Home*. UNICEF, 21 Mar. 2005. Web. 23 Oct. 2011. http://www.unicef.org/wash/index_25637.html.

Wang, Hao, and Jing Liu. "Mobile Phone Based Health Care Technology." *Recent Patents on Biomedical Engineering* 2.1 (2010): 15–21. Print.

Wang, S., Blackford Middleton, et al. "A Cost-benefit Analysis of Electronic Medical Records in Primary Care." *The American Journal of Medicine* 114.5 (2003): 397–403. Print.

Wernsdorfer, Walther, Simon Hay, and Dennis Shanks. "Learning from History." *Shrinking the Malaria Map: a Prospectus on Malaria Elimination*. By Richard G. Feachem, Allison A. Phillips, and G. A. T. Targett. San Francisco: Global Health Group, UCSF Global Health Sciences, 2009. 95–107. Print.

West, Darrell M., and Edward Alan. Miller. *Digital Medicine: Health Care in the Internet Era*. Washington, D.C.: Brookings Institution, 2009. Print.

Zundel, K. M. "Telemedicine: History, Applications, and Impact on Librarianship." *PubMed Central* 84.1 (1996): 71–79. Print.

CHAPTER 8: EDUCATION

2012 World Hunger and Poverty Facts and Statistics. World Hunger Education Service, 2012. Web. 13 Feb. 2012. http://www.worldhunger.org/articles/Learn/world%20hunger%20facts%202002.htm#Number_of_hungry_people_in_the_world.

"A History of College Inflation." *College Money*. Marlton, NJ 2004-2006. Web. 17 Apr. 2011. http://www.collegemoney.com/images/News/News_12_4.pdf.

"Africa Education Watch: Good Governance Lessons for Primary Education." *Transparency.org*. Transparency International, 23 Feb. 2010. Web. http://www.transparency.org/news_room/in_focus/2010/african_education_watch.

"AIML—The Artificial Intelligence Markup Language." *A.L.I.C.E. Artificial Intelligence Foundation*. A.L.I.C.E. AI Foundation, Inc. Web. 17 Apr. 2011. http://www.alicebot.org/aiml.html.

Alismail et al., "Combining Web Technology and Mobile Phones to Enhance English Literacy in Underserved Communities." *First Annual Symposium on Computing for Development (ACM DEV 2010)*. 17–18 Dec. 2010, London, UK.

"All the (Synthetic) World's a Stage." *Humanities, Then and Now* Fall XXIX.1 (2006). *Indiana.edu*. Indiana University. Web. 17 Apr. 2011. http://www.indiana.edu/~rcapub/v29n1/synthetic.shtml.

Archibald, Robert B., and David H. Feldman. *Why Does College Cost so Much?* Oxford: Oxford UP, 2011. Print.

Auguste, Byron, Paul Kihn, and Matt Miller. "Closing the Talent Gap: Attracting and Retaining Top-Third Graduates to Careers in Teaching." *McKinsey On Society*. McKinsey, Sept. 2010. Web. 23 Oct. 2011. http://mckinseyonsociety.com/downloads/reports/Education/Closing_the_talent_gap.pdf.

Bosner, Ulrich. "Return on Educational Investment." *Americanprogress.org*. Center for American Progress, 19 Jan. 2011. Web. 23 Oct. 2011. http://www.americanprogress.org/issues/2011/01/educational_productivity/report.html.

Burtless, Gary. *Does Money Matter? The Effect of School Resources on Student Achievement*. Washington D.C.: The Brookings Institute, 1996.

Busch, Fritz. "IPads Preferred to Books at GFW." *NUJournal.com*. The Journal, 2 Mar. 2011. Web. http://www.nujournal.com/page/content.detail/id/523011.html.

"Casino at Resorts World Sentosa, Singapore—Casino Levy." *Resorts World Sentosa | Singapore Hotel & Casino Resort | Integrated Resort*. Resorts World. Web. 7 Nov. 2011. http://www.rwsentosa.com/language/en-US/Gaming/CasinoLevy.

Chaudhary, Latika. "Determinants of Primary Schooling in British India." Hoover Institution, Stanford University, 2 Aug. 2007.

Chen, Shaohua, and Martin Ravallion. *The Developing World is Poorer Than We Thought, but No Less Successful in the Fight Against Poverty*. Washington, DC: World Bank, 2008. Print.

"College Prices Increase in Step with Inflation." *College Board—News & Press*. 29 Oct. 2008. Web. 17 Apr. 2011. http://press.collegeboard.org/releases/2008/college-prices-increase-step-inflation.

"Corruption in Education System in India." *Www.Azadindia.org*. AZAD India Foundation, 2010. Web. http://www.azadindia.org/social-issues/education-system-in-india.html.

"CPS October 2009—Detailed Tables—U.S Census Bureau." *Census Bureau Home Page*. US Census Bureau, Oct. 2009. Web. Oct. 2011. http://www.census.gov/hhes/school/data/cps/2009/tables.html.

D'Orio, Wayne. "IPads in Class | Scholastic.com." *Scholastic | Children's Books and Book Club | Scholastic.com*. Scholastic. Web. http://www.scholastic.com/browse/article.jsp?id=3755865.

"Editorial: Civil and Political Rights in Schools." CRIN—Child Rights Information Network—Education. Child Rights International Network, 2011. Web. Nov. 2011. http://www.crin.org/themes/ViewTheme.asp?id=7.

Edusim—3D Virtual Worlds for the Classroom Interactive Whiteboard. Web. http://edusim3d.com/.

"Facts and Figures: 2010–2011." *Public Schools of North Carolina*. Web. 17 Apr. 2011. http://www.ncpublicschools.org/docs/fbs/resources/data/factsfigures/2010-11 figures.pdf.

Fischer, Kurt W. "Mind, Brain, and Education: Building a Scientific Groundwork for Learning and Teaching." *Mind, Brain, and Education* 3.1 (2009): 3–16. Print.

"GDP—per Capita (PPP)—Country Comparison." *Index Mundi—Country Facts*. Index Mundi, 2011. Web. 7 Nov. 2011. http://www.indexmundi.com/g/r.aspx?v=67.

Glaeser, Edward. *Triumph of the City*. New York: Penguin, 2011.

"Goddard Enters the World of Online Gaming." *GSFC News: Tech Transfer* Winter 6.1 (2007/08): 2. *Fuentek.net*. National Aeronautics and Space Administration. Web. 17 Apr. 2011. http://fuentek.net/downloads/gsfc-ttnews_winter08.pdf.

Hanushek, Eric. *Schoolhouses, Courthouses, and Statehouses: Solving the Funding-Achievement Puzzle in America's Public Schools*. Princeton, NJ: Princeton University Press, 2009.

Herbst, Jurgen. *The Once and Future School: Three Hundred and Fifty Years of American Secondary Education*. New York: Routledge, 1996. Print.

Hinton, Christina, Koji Miyamoto, and Bruno Della-Chiesa. "Brain Research, Learning and Emotions: Implications for Education Research, Policy and Practice." *European Journal of Education* 43.1 (2008): 87–103. Print.

Huebler, Friedrich. "Adult Literacy Rates." *International Education Statistics*. 8 July 2007. Web. http://huebler.blogspot.com/2007/07/adult-literacy-rates.html.

Kim Hwa-young, Theresa. "Coming Soon "Digital Schools" *AsiaNews.it*. PIME, 9 Mar. 2007. Web. http://www.asianews.it/index.php?l=en&art=8678.

Luce, Edward. *In Spite of the Gods: the Strange Rise of Modern India*. New York: Doubleday, 2007. Print.

MacWilliams, Byron. "In Georgia, Professors Hand Out Price Lists." *The Chronicle of Higher Education*. N.p., 2 Aug. 2002. Web. http://chronicle.com/article/In-Georgia -Professors-Hand/33565.

McCurry, Justin. "Mobile Phone Exam Cheat Shocks Japanese Meritocracy | World News | The Guardian." *Latest News, Comment and Reviews from the Guardian | Guardian.co.uk*. 4 Mar. 2011. Web. 17 Apr. 2011. http://www.guardian.co.uk/ world/2011/mar/04/japan-mobile-phone-exam-cheat.

Najibullah, Farangis. "In Central Asia, Corruption Undermining Education System." *Radio Free Europe / Radio Liberty—Free Media in Unfree Societies*. Radio Free Europe Radio Liberty, 6 Aug. 2009. Web. http://www.rferl.org/content/In_Central_Asia _Bribery_A_Common_Part_of_Education/1794065.html.

OECD (2011), Lessons from PISA for the United States, Strong Performers and Successful Reformers in Education, OECD Publishing. http://dx.doi.org/10.1787/ 9789264096660-en

Open Wonderland. Web. 17 Apr. 2011. http://openwonderland.org/

"Outcomes for College Graduates." *U.S. Department of Education*. U.S. Department of Education, Jan. 1999. Web. 23 Oct. 2011. http://www2.ed.gov/pubs/College ForAll/graduates.html.

Patel, Ila. *Education for All—Mid Decade Assessment: Adult Literacy and Lifelong*

Learning in India. New Delhi: National University of Educational Planning and Administration, 2009. Web. http://www.educationforallinindia.com/Adult-Literacy -and-Lifelong-Learning-in-India.pdf.

"Princeton University | Fees & Payment Options." *Princeton University Undergraduate Admissions.* Princeton University, 2012. Web. 5 Mar. 2012. http://www.prince ton.edu/admission/financialaid/cost/.

Patrinos, Harry A., and George Psacharopoulos. "Education Past, Present and Future Global Challenges." (2011): 15–24. The World Bank Human Development Network Education Team. Web. http://www-wds.worldbank.org/servlet/WDSContentServer/ WDSP/IB/2011/03/29/000158349_20110329095336/Rendered/PDF/WPS5616.pdf.

"Per Capita Personal Income." *Infoplease: Encyclopedia, Almanac, Atlas, Biographies, Dictionary, Thesaurus.—Infoplease.com.* Pearson Education, 2011. Web. 7 Nov. 2011. http://www.infoplease.com/ipa/A0104547.html.

Prince, Michael. "Does Active Learning Work? A Review of the Research." *Journal of Engineering Education* 93.3 July (2004): 223–231. Print.

"Revenues and Expenditures for Public Elementary and Secondary Education—Selected Findings: Fiscal Year 2007." *National Center for Education Statistics (NCES) Home Page, a Part of the U.S. Department of Education.* Institute of Educational Services, Mar. 2009. Web. 14 Feb. 2012. http://nces.ed.gov/pubs2009/expenditures/ findings.asp.

Schemo, Diana J. "Failing Schools Strain to Meet U.S. Standard." *NYTimes.com.* New York Times, 16 Oct. 2007. Web. 17 Apr. 2011. http://www.nytimes.com/2007/10/16/ education/16child.html?pagewanted=print.

"Schools Lose Records; English Learners Pay." *PhysOrg.com—Science News, Technology, Physics, Nanotechnology, Space Science, Earth Science, Medicine.* 11 Apr. 2011. Web. 17 Apr. 2011. http://www.physorg.com/news/2011-04-schools-english -learners.html.

Shaw, Philip. "Educational Corruption and Growth." University of Connecticut, Nov 12, 2007, working paper. Web. http://www.econ.yale.edu/seminars/macro/mac09/ shaw-090409.pdf.

"Statistics Singapore—Key Annual Indicators." *Department of Statistics Singapore.* Singapore Government, 2011. Web. 7 Nov. 2011. http://www.singstat.gov.sg/stats/ keyind.html.

"Statistics Singapore—Time Series on Population." *Department of Statistics Singapore.* Singapore Government, 2011. Web. 7 Nov. 2011. http://www.singstat.gov.sg/ stats/themes/people/hist/popn.html.

"UNICEF—Cuba—Statistics." *UNICEF.org.* UNICEF, 2 Mar. 2010. Web. 7 Nov. 2011. http://www.unicef.org/infobycountry/cuba_statistics.html.

Voss, Joel L., Brian D. Gonsalves, Kara D. Federmeier, Daniel Tranel, and Neal J. Cohen. "Hippocampal Brain-network Coordination during Volitional Exploratory Behavior Enhances Learning." *Nature Neuroscience* (2010): 115–120. Print.

Wolfgang, Ben. "S. Korea Leads Way for Paperless Classroom." *The Washington Times.* The Washington Times, 18 July 2011. Web. 7 Nov. 2011. http://www.washington times.com/news/2011/jul/18/s-korea-leads-way-for-paperless-classroom/?page=all.

Yazzie-Mintz, Ethan. "Charting the Path from Engagement to Achievement: A Report on the 2009 High School Survey of Student Engagement." *Indiana.edu.* Indiana University, 2010. Web. 17 Apr. 2011. http://www.indiana.edu/~ceep/hssse/images/HSSSE_2010_Report.pdf.

CHAPTER 9: DEVELOPING WORLD

Aker, Jenny, and Isaac Mbiti. "Africa Calling Can Mobile Phones Make a Miracle?" *Boston Review.* Boston Review, Mar/Apr. 2010. Web. http://bostonreview.net/BR35.2/aker_mbiti.php.

Aker, Jenny, and Isaac Mbiti. "Mobile Phones and Economic Development in Africa." *Journal of Economic Perspectives* 24.3 (2010): 207–232. Print.

Andersen, Thomas B., Jeanet Bentzen, Carl-Johan Dalgaard, and Pablo Selaya. "Does the Internet Reduce Corruption? Evidence from U.S. States and Across Countries," *The World Bank Economic Review.* 15Apr. 2010.

Apambire, W. Braimah. "Water Supplies in Developing Countries, The West Africa Water Initiative as an Example." The Water Center Seminar. University of Washington. Seattle. 27 Feb. 2007. Presentation.

Banerjee, Abhijit V., and Esther Duflo. "Giving Credit Where It Is Due." *Journal of Economic Perspectives* 24.3 (2010): 6180. Print.

Bridges, Laurie, Hannah Gascho Rempel, and Kimberly Griggs. "Making the Case for a Fully Mobile Library Web Site: From Floor Maps to the Catalog." *Reference Services Review* 38.2 (2010). Print.

Chaia, Alberto, Tony Goland, and Robert Schiff. "Counting Africa's Unbanked." *McKinsey on Africa: A Continent on the Move.* June 2010, pp. 66–67.

"China GDP Annual Growth Rate." *TradingEconomics.com—Economic Data for 196 Countries.* Trading Economics, 2011. Web. 07 Nov. 2011. http://www.tradingeconomics.com/china/gdp-growth-annual.

Easterly, William. *The White Man's Burden.* New York: Penguin, 2007. Print.

"Electric Power Consumption (kWh per Capita) | Data | Table." *Data | The World Bank.* The World Bank, 2011. Web. 07 Nov. 2011. http://data.worldbank.org/indicator/EG.USE.ELEC.KH.PC.

Esipisu, Isaiah. "Kenyan Farmers Use SMS to Beat Climate-driven Price Uncertainty." *Trust.org.* AlertNet: The World's Humanitarian News Site, 18 Apr. 2011. Web. www.trust.org/alertnet/news/kenyan-farmers-use-sms-to-beat-climate-driven-price-uncertainty/.

Faculty of Economics, University of Groningen. "Chinese Economic Performance in the Long-Run." *Ggdc.net.* OECD Development Centre, Paris, 1998. Web. http://www.ggdc.net/maddison/China_book/Chap_2_tables/Table2.2ab.pdf.

Gandhi, Vasant P., and N.V. Namboodiri. "Marketing of Fruits and Vegetables in India: A Study Covering the Ahmedabad, Chennai and Kolkata Markets." *www.iimahd.ernet.* N.p., 9 June 2004. Web. 25 Oct. 2011. www.iimahd.ernet.in/publications/data/2004-06-09vpgandhi.pdf.

"Global Mobile Statistics 2011: All Quality Mobile Marketing Research, Mobile Web

Stats, Subscribers, Ad Revenue, Usage, Trends . . . | MobiThinking." *Home | Mobi-Thinking*. Web. 15 Oct. 2011. http://mobithinking.com/mobile-marketing-tools/latest-mobile-stats.

Goel, Rajeev K., and Michael A. Nelson. *Causes of Corruption History, Geography, and Government*. Helsinki: Bank of Finland, Institute for Economies in Transition, 2008. Web. http://www.suomenpankki.fi/bofit_en/tutkimus/tutkimusjulkaisut/dp/Documents/DP0608.pdf?hl=Causes%20of%20Corruption:%20History,%20Geography%20and%20Government.

Grönlund, Åke, Rebekah Heacock, David Sasaki, Johan Hellström, and Walid Al-Saqaf. *Increasing Transparency and Fighting Corruption Through ICT: Empowering People and Communities*. Sweden: Spider, 2010. Print.

Jack, William, and Tavneet Suri. *The Economics of M-PESA*. MIT: Working Paper, 2010. Web. http://www.mit.edu/~tavneet/M-PESA.pdf.

Jensen, Robert. "The Digital Provide: Information (Technology), Market Performance, and Welfare in the South Indian Fisheries Sector." *The Quarterly Journal of Economics* 122.3 (2007): 879–924. Print

Kurata, Phillip. "Mobile Phones Connect Ugandan Farmers to Agricultural Information," *America.gov Archive*, 20 Sept. 2010. Web. www.america.gov/st/develop-english/2010/September/20100920153017cpataruk0.4970819.html?CP.rss=true.

Lievrouw, Leah A., and Sonia M. Livingstone. *Handbook of New Media: Social Shaping and Social Consequences of ICTs*. London: SAGE, 2006. Print. p. 321.

Manjaro, Nadine. "International Markets Remain Untapped Potential for Tower Companies," *RCR Wireless*. RCR Wireless, 5 Mar. 2010. Web. www.rcrwireless.com/ARTICLE/20100305/25_YEARS/100309967/international-markets-remain-untapped-potential-for-tower-companies.

Mittal, Surabhi, Sanjay Gandhi, and Gaurav Tripathi. *Socio-economic Impact of Mobile Phones on Indian Agriculture: Working Paper No. 246*. New York: Indian Council For Research On International Economic Relations, 2010. Print.

"Mobile Services in Poor Countries: Not Just Talk." *The Economist*. The Economist, 27 Jan. 2011. Web. 25 Oct. 2011. http://www.economist.com/node/18008202.

Narula, Sapna A., and Navin Nainwal. "ICTs and Agricultural Supply Chains Opportunities and Strategies for Successful Implementation." *Information Technology in Developing Countries* 20.1 (2010): Article 7. *Information Technology in Developing Countries*. Web. http://www.iimahd.ernet.in/egov/ifip/feb2010/sapna-narula.htm.

Nichter, Simeon, and Lara Goldmark. "Small Firm Growth in Developing Countries." *World Development* 37.9 (2009): 1453-1464. Print.

"Number of Credit Card Holders Slips to 18.3 Million." *India Business News, Stock Market, Personal Finance, Economy—Rediff.com*. ediff.com, 18 May 2010. Web. 25 Oct. 2011. http://business.rediff.com/report/2010/may/18/number-of-credit-card-holders-slips-to-18-point-3-million-in-march.htm.

O'Brien, Kevin J. "Nokia Taking a Rural Road to Growth." *Nytimes.com*. New York Times, 1 Nov. 2010. Web. 7 Nov. 2011. http://www.nytimes.com/2010/11/02/technology/02nokia.html.

Poisson, Muriel. *Corruption and Education*. Paris: UNESCO, 2010.

Rosenberg, Tina. "Doing More than Praying for Rain." *NYTimes.com—Opinion Pages*. The New York Times, 9 May 2011. Web. http://opinionator.blogs.nytimes.com/2011/05/09/doing-more-than-praying-for-rain/?partner=rss&emc=rss.

"Sanitation and Hygiene." *Water.worldbank.org*. The World Bank, 2011. Web. 7 Nov. 2011. http://water.worldbank.org/water/topics/sanitation-and-hygiene.

Stepanek, Marcia. "New Social Media Trend: micro-multinationals." *Justmeans Business. Better.* Justmeans, 26 Apr. 2010. Web. http://www.justmeans.com/New-social-media-trend-micro-multinationals/13602.html.

Sullivan, Kevin. "For India's Traditional Fishermen, Cellphones Deliver a Sea Change." *Washington Post*, 15 Oct. 2006. Web. http://www.washingtonpost.com/wp-dyn/content/article/2006/10/14/AR2006101400342.html.

"The Road to Hell Is Unpaved." *The Economist*. The Economist, 19 Dec. 2002. Web. 25 Oct. 2011. http://www.economist.com/node/1487583?Story_ID=1487583.

"The World Factbook: GDP—Composition by Sector." *Cia.gov*. Central Intelligence Agency. Web. 7 Nov. 2011. https://www.cia.gov/library/publications/the-world-factbook/fields/2012.html.

"The World Factbook: Labor Force—By Occupation." *Cia.gov*. Central Intelligence Agency. Web. 7 Nov. 2011. https://www.cia.gov/library/publications/the-world-factbook/fields/2048.html.

van Meel, Juriaan, and Paul Vos. "Funky Offices: Reflections on Office Design in the 'New Economy.'" *Journal of Corporate Real Estate* 3.4 (2001): 322–334. Print.

Yunus, Muhammad. *Creating a World Without Poverty*. New York: Public Affairs, 2007.

CHAPTER 10: NEW WORLD

"A History of American Agriculture: Farmers & the Land." Growing a Nation: The Story of American Agriculture. Agclassroom.org. Web. http://www.agclassroom.org/gan/timeline/farmers_land.htm.

"Building with Big Data." *The Economist—World News, Politics, Economics, Business & Finance*. The Economist, 26 May 2011. Web. 25 Oct. 2011. http://www.economist.com/node/18741392.

Cisco Visual Networking Index: Global Mobile Data Traffic Forecast Update, 2010–2015. Rep. San Jose, CA: Cisco, 2011.

Elliott, Justin. "The 10 Most Important WikiLeaks Revelations." *Salon.com*. Salon Media Group, 29 Nov. 2010. Web. 15 Oct. 2011. http://www.salon.com/news/politics/war_room/2010/11/29/wikileaks_roundup.

Hewitt, Bill. "Big Data: Big Costs, Big Risks And Big Opportunity." *Information for the World's Business Leaders—Forbes.com*. Forbes, 27 May 2011. Web. 13 Oct. 2011. http://blogs.forbes.com/ciocentral/2011/05/27/big-data-big-costs-big-risks-and-big-opportunity/.

Lawler, Andrew. "Saving Iraq's Treasures." The Smithsonian, 1 June 2003. Web. 1 Nov. 2011. http://www.andrewlawler.com/smithsonian/item/8-saving-iraqs-treasures-cover-june-2003.html.

Lucas, Robert E. "The Industrial Revolution: Past and Future: 2003 Annual Report Essay."

minneapolisfed.org. The Federal Reserve Bank of Minneapolis, 1 May 2004. Web. 7 Oct. 2011. http://www.minneapolisfed.org/publications_papers/pub_display.cfm?id =3333.

Manyika, James, et al. *Big Data: The Next Frontier for Innovation, Competition, and Productivity*. Sydney: McKinsey Global Institute, 2011. Print.

Pélissié du Rausas, Matthieu, et al. *Internet Matters: The Net's Sweeping Impact on Growth, Jobs, and Prosperity*. New York: McKinsey & Company, 2011. Web. 7 Oct. 2011. http://www.mckinsey.com/mgi/publications/internet_matters/pdfs/MGI_inter net_matters_full_report.pdf.

Razin, Assaf, and Efraim Sadka. *Population Economics*. Cambridge, MA: MIT, 1995. Print.

"Rural Labor and Education: Farm Labor." USDA Economic Research Service—Home Page. United States Department of Agriculture, 11 July 2011. Web. http://www.ers.usda.gov/ Briefing/LaborAndEducation/FarmLabor.htm.

Sheppard, Brett. "Putting Big Data to Work: Opportunities for Enterprises." *GigaOM Pro*. GigaOM, 22 Mar. 2011. Web. 13 Oct. 2011. http://pro.gigaom.com/ 2011/03/putting-big-data-to-work-opportunities-for-enterprises/.

"The American Workplace—The Shift to a Service Economy." StateUniversity.com. Web. http://jobs.stateuniversity.com/pages/16/American-Workplace-SHIFT-SERVICE -ECONOMY.html.

Warren, Samuel, and Louis Brandeis. "The Right to Privacy." *Harvard Law Review* 4.193 (1890): 196. Print.

Watters, Audrey. "Gnip CEO on the Challenges of Handling the Real-Time, Big Data Firehose." *ReadWriteCloud*. ReadWriteWeb, 25 May 2011. Web. 14 Oct. 2011. http://www.readwriteweb.com/cloud/2011/05/gnip-ceo-on-the-challenges-of.php.

"World FactBook." *Central Intelligence Agency Publications*. Central Intelligence Agency, 12 Oct. 2011. Web. 12 Oct. 2011. https://www.cia.gov/library/publications/ the-world-factbook/geos/us.html.

Zeitlin, June. "Data on the Status of Women Worldwide." WEDO. Women's Environment & Development Organization, 22 May 2006. Web. 01 Nov. 2011. http://www.wedo.org/library/data-on-the-status-of-women-worldwide.

ACKNOWLEDGMENTS

I would like to thank Dan McNeill, Paul Freiberger, my MicroStrategy team of Mark LaRow, Kevin Spurway, Kirsten Brown, Rachel Blum, Lou Martinage, Warren Getler, and Belinda Morrissette, and my Personal Assistant, Rodney Richardson, for their help in the making of this book. Without their efforts this book would not have been possible.

I N D E X